LABORATORY MANUAL
to Accompany

Introductory Electric Circuits:
Electron Flow Version

Introductory Electric Circuits:
Conventional Flow Version

William Muckler

With Editorial Assistance by Robert Paynter

Prentice Hall

Upper Saddle River, New Jersey Columbus, Ohio

Cover Photo: © FPG International
Editor: Linda R. Ludewig
Production Editor: Rex Davidson
Design Coordinator: Karrie M. Converse
Cover Designer: Ceri Fitzgerald
Production Manager: Patricia A. Tonneman
Marketing Manager: Ben Leonard

This book was set in Times Roman by The Clarinda Company and was printed and bound by Courier/Kendallville. The cover was printed by Phoenix Color Corp.

Simon & Schuster/A Viacom Company
Upper Saddle River, New Jersey 07458

Printed in the United States of America

10 9 8 7 6 5 4 3 2 1

ISBN: 0-02-392501-9

Prentice-Hall International (UK) Limited, *London*
Prentice-Hall of Australia Pty. Limited, *Sydney*
Prentice-Hall of Canada, Inc., *Toronto*
Prentice-Hall Hispanoamericana, S. A., *Mexico*
Prentice-Hall of India Private Limited, *New Delhi*
Prentice-Hall of Japan, Inc., *Tokyo*
Simon & Schuster Asia Pte. Ltd., *Singapore*
Editora Prentice-Hall do Brasil, Ltda., *Rio de Janeiro*

Contents

EWB Exercises

The following is a list of the exercises contained on the EWB disk for this lab manual. The EWB exercises were developed by John Reeder, Merced Community College, Ca.

1_1 Using the DMM found in Electronics Workbench

1_2 Using the voltmeter and the ammeter found in EWB

1_3 EWB component symbols

1_4 The basic circuit: Batteries, lamps, switches, and fuses

2_1 Determining resistor color codes using resistor values and tolerances

2_2 Using the DMM to measure resistance in EWB

2_3 Resistance and conductance as reciprocals

3_1 Setting potentiometer resistance in EWB

4_1 Constructing basic circuits

5_1 Total resistance in a series circuit

6_1 Using the DMM to measure current in a series circuit

6_2 Calculating and measuring current in a series circuit

7_1 Calculating current using Ohm's law

7_2 Calculating resistance using Ohm's law

7_3 Calculating voltage using Ohm's law

7_4 Determining voltage, current, and resistance using Ohm's law

7_5 Interaction of voltage, current, and resistance in a basic circuit

Preface

This lab manual has been written to accompany the textbook *Introductory Electric Circuits* by Robert Paynter. The exercises in this manual have been arranged to follow as closely as possible the material covered in the textbook.

The exercises in this manual have been tested in our lab. We recognize the fact that there is no such thing as a universal lab, but an attempt has been made to present these exercises so that the student has some discretion on how to proceed through each. We hope this presentation allows the student to develop a sense of self-discovery in learning a method into investigating electric circuits. **Note that the current arrows in this manual are drawn according to the electron flow format.**

Acknowledgments

This lab manual would not have been written by me if Robert Paynter hadn't given me the opportunity to do what I have wanted to do for 25 years. I also want to give my deepest thanks to my daughter Catherine Miller, who not only read this stuff but actually did the preliminary editing of the manuscript and was very supportive. I thank my wife, who kept me fed and supported my efforts.

William Muckler

Parts List

EQUIPMENT

Dual-trace oscilloscope
Audio function generator
Dual-polarity dc power supply
Test leads (3 sets)
1.5 V D cells (2)
9 V battery
SPST switch
Digital multimeter (DMM)
Volt-ohm-milliammeter (VOM)
Protoboard or breadboard
Decade resistance box
Hookup wires
12.6 Vac transformer
D-cell battery holders (2)

POTENTIOMETERS (0.5 W)

1 kΩ linear taper
10 kΩ linear taper
15 kΩ linear taper
10 kΩ audio taper

RESISTORS (0.5 W, UNLESS NOTED OTHERWISE)

10 Ω (3)
22 Ω
33 Ω
47 Ω
56 Ω
68 Ω
100 Ω
120 Ω
122 Ω
150 Ω

150 Ω
200 Ω
220 Ω
330 Ω, 0.25 W
330 Ω (2)
470 Ω
560 Ω
1 kΩ
1.2 kΩ
1.5 kΩ (2)

1.8 kΩ
2 kΩ
2.2 kΩ
3.3 kΩ
4.7 kΩ
10 kΩ
15 kΩ
22 kΩ
1.5 MΩ

CAPACITORS (50 V_{DC})

10 μf
1 μf
0.1 μf
0.001 μf
3.3 nf

INDUCTORS

10 mH
1 mH

Part 1

dc Exercises

Exercise 1

Equipment Familiarization

OBJECTIVES

After completing this exercise, you should be able to:

1. Locate the safety equipment and test equipment in your lab.
2. Identify the following: the digital multimeter (DMM), volt-ohm-milliammeter (VOM), and dc power supply.
3. Locate any components and parts needed to complete an exercise.
4. Describe the function of the DMM, VOM, and dc power supply.
5. Measure the output voltage from the dc power supply using the DMM and/or VOM.

LAB PREPARATION

Review Section 1.2 of *Introductory Electric Circuits.*

DISCUSSION

Throughout your studies in electricity and electronics, you will be introduced to various types of test equipment. One purpose of this exercise is to introduce you to several of the most basic test instruments: the digital multimeter (DMM), the volt-ohm-milliammeter (VOM), and the dc power supply.

The Digital Multimeter (DMM)

The DMM can be used to measure voltage (V), current (I), and resistance (R). A typical DMM is shown in Figure 1.1. As you can see, the DMM has a numeric readout. (This feature distinguishes the DMM from the VOM, which typically has a needle that moves over a fixed scale.)

The DMM is set up to measure voltage as follows:

1. Connect the test leads to the input jacks of the meter, red to positive and black to ground.
2. The *function control* tells the meter the type of value measured. The function control is set for a dc voltage by either
   a. Pushing the dc volts button.
   b. Turning the selector switch (on some models) to the dc volts position.
3. The range selector tells the meter the approximate range of the values for the measurement. The meter range is selected by either
   a. Pushing the range select button for a setting that is greater than the value you expect to read.
   b. Turning the range selector switch (on some models) to a setting that is greater than the value you expect to read.

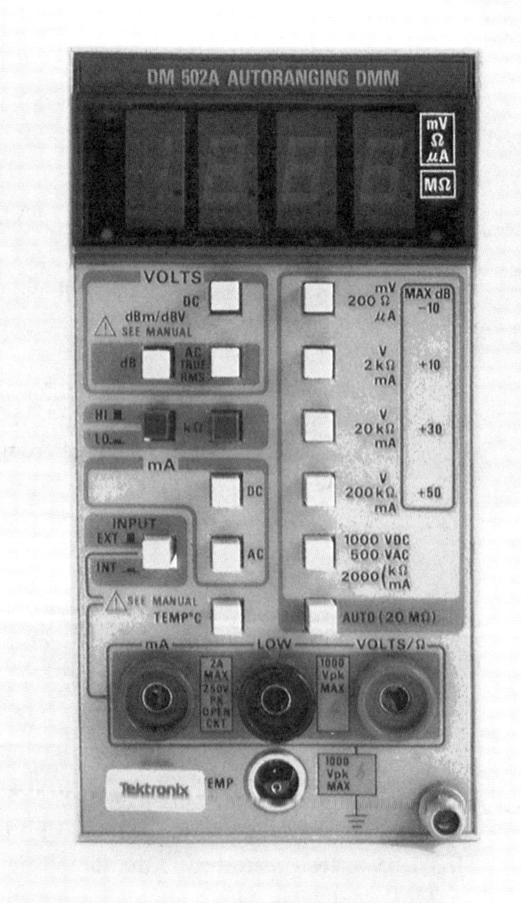

FIGURE 1.1 Digital multimeter.

For example, if you know a voltage will be somewhere around 10 V, the meter is set to the lowest range that is greater than 10 V.

Several important points need to be made at this time:

1. Your DMM may use labels that are different than those presented here. Your instructor can help you make the needed adjustments to the setup procedure.
2. *When you are measuring an unknown voltage, the range selector is always set to the highest range. Then, after the test probes are connected to the circuit, the range is slowly decreased until a usable reading is obtained.*
3. In most cases, a DMM readout will flash if the selected range is too low for the value measured. In this case, the setting is increased (in steps) until the readout stops flashing.

In Exercises 2 and 6, you will be shown how to measure resistance and current.

The Volt-Ohm-Milliammeter (VOM)

A DMM is often referred to as a multimeter, which means it can be used to measure various circuit values. Another type of multimeter is the volt-ohm-milliammeter, or VOM.

VOMs have a needle that moves over a fixed scale, as shown in Figure 1.2. The position of the needle on the scale indicates the value of the measurement. In general, VOM voltage readings are less accurate than DMM voltage measurements for two reasons:

1. Many analog meters are subject to parallax error. Parallax error is caused by reading the meter from an angle. (Parallax error can be demonstrated by reading a clock from the extreme left or right side.)

FIGURE 1.2 Volt-ohm-milliammeter (VOM).

2. Every meter has an input resistance rating. The lower the input resistance of a multimeter, the less accurate its voltage reading. In most cases, VOMs have lower input resistance ratings than DMMs. As a result, VOM voltage readings are often less accurate than DMM voltage readings.

Since DMMs provide more accurate voltage readings, they are used more often than VOMs. However, there will be times when you need to use a VOM, or it might be the only meter available. Thus, you need to know how to measure voltage using a VOM. A VOM is set for a voltage reading measurement as follows:

1. Connect the red test lead to the + input jack, and the black test lead to the − or common input jack.
2. Set the −dc/+dc switch to
   a. +dc if you are measuring positive dc voltages.
   b. −dc if you are measuring negative dc voltages.
3. Set the range selector to the proper setting.

> Note: If the range setting on the VOM is too low, the voltage measured may "peg the meter" (slam the needle to the high end of the scale). In some cases, this can damage the meter movement. To protect the meter, always start a VOM in its highest scale and slowly decrease the setting until a midscale reading can be obtained.

4. Ground the input test leads together, then adjust the zero adjust knob so that the needle is directly over the 0 V position on the meter face. *Note:* Every time that you change the range selector, you must reset the zero adjust knob.

Again, your VOM may use labels that are different than those presented here. Your instructor can help you make the needed adjustments to the setup procedure.

The dc Power Supply

The dc power supply is a piece of equipment that provides a dc output voltage that can be adjusted over a range of values. As such, it essentially acts as a variable dc voltage source. The basic outputs and controls for a variable dc power supply are shown in Figure 1.3.

Most of the settings for the dc power supply are made before the device is turned on. Before turning on your dc power supply:

1. Make sure that both voltage control knobs (+ volts and − volts) are turned fully counterclockwise. This step sets your output voltage at zero. If the power supply has dual-tracking capabilities, make sure that the dual-tracking control is disabled.
2. If your dc power supply has current limit controls, make sure they are set for maximum possible output current.

> Note: If your dc power supply has a current limit light, it should remain off. If a current limit light comes on, turn off your dc power supply and notify your instructor.

3. Connect a black test lead to the common output and a red test lead to the appropriate volts output (+V or −V on many dc power supplies).

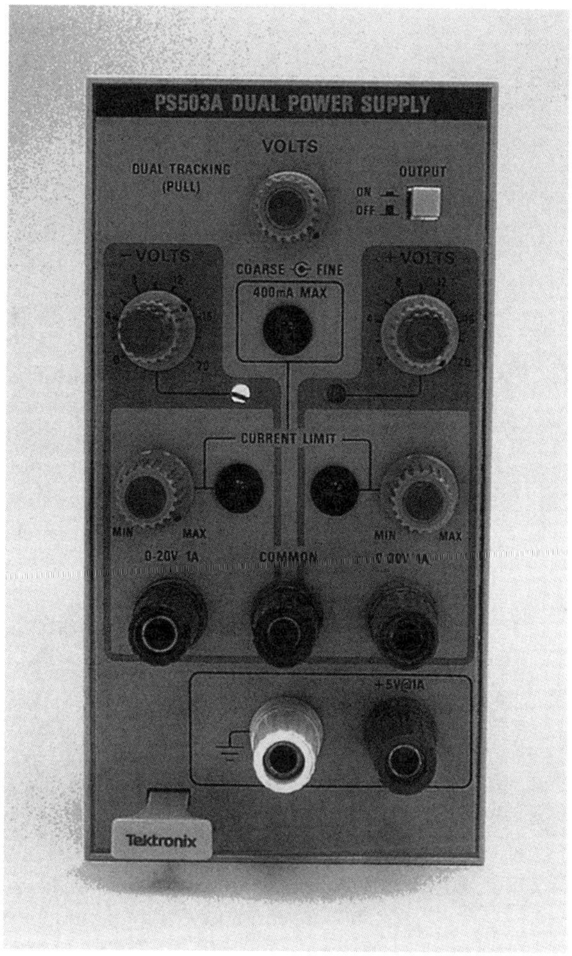

FIGURE 1.3 Typical dc power supply.

4. Connect the other ends of the test leads to the connection points in the circuit being tested.

Once the dc power supply is set up and connected to the circuit, the output voltage from the supply is set as follows:

1. Connect the DMM to the output from the power supply: positive to positive, negative to negative.
2. Make sure that both the DMM and the dc power supply are turned on.
3. Slowly turn the volts control knob clockwise until you read the desired setting on the DMM.

As you practice using the DMM and the dc power supply, you'll see that it takes very little time to set up a circuit for an exercise.

MATERIALS

1 dc power supply
1 DMM
1 VOM
4 test leads (two red and two black)

PROCEDURE

1. Locate the electrical circuit breakers in the lab. In case of an emergency, these breakers can be used to disconnect all the power to the room.
2. Locate the fire extinguisher.
3. Locate the test leads. Take two red and two black test leads to your lab station.
4. Locate the hookup wires and the resistors.
5. Locate the protoboard or the breadboard that you will be using during your exercise.
6. Locate the DMM and dc power supply at your station. Compare them and identify the different controls for each. Have your instructor help if you have problems.
7. Using the procedure outlined earlier, set your DMM to read dc volts. (Select the lowest range that is greater than 10 V_{dc}.)
8. Using the procedure outlined earlier, perform the pre-power setup for the dc power supply. The supply leads should be connected for a positive dc output voltage.
9. If available, set the VOM to read positive voltage. Set the VOM range to the 50 V_{dc} (or similar value) scale. Be sure that you set the zero-adjust knob.
10. Connect the DMM to the output of the dc power supply. Apply power to the dc supply and adjust its output (as close as you can) to 10 V_{dc}. Record the actual DMM reading below.

    $V =$ ____________ (DMM)
11. Replace the DMM with the VOM. Record (as accurately as possible) the VOM reading below.

    $V =$ ____________ (VOM)
12. Repeat Steps 10 and 11, this time setting the output from the dc power supply for 15.5 V_{dc}. Record the readings (as accurately as possible) below.

    $V =$ ____________ (DMM)

    $V =$ ____________ (VOM)
13. Turn off the dc power supply and the meter(s). Return all equipment and leads to their proper locations.

QUESTIONS/PROBLEMS

1. What quantities can be measured using a typical digital multimeter (DMM)?

   a. ____________

   b. ____________

   c. ____________

2. What purpose is served by each of the following multimeter controls?
   a. The function selector.

   b. The range selector.

3. What is parallax error? What type of meter may experience this type of error?

4. Which type of multimeter usually provides more accurate voltage readings?

5. What can happen if the range setting on a VOM is too low?

6. Precision is a measure of how exact a value is. For example, a temperature reading of 25.645° C is more precise than a reading of approximately 25.6° C. With this in mind, compare the DMM and VOM readings in Steps 10 and 11.

   Which do you believe to be the most precise reading? Why? ________________

   ____

   ____

7. For the readings in Step 12, which do you believe to be the most precise reading? Why? ____

   ____

Exercise 2

Resistor Color Code and the Ohmmeter

OBJECTIVES

After completing this exercise, you should be able to:

1. Determine the nominal value of a resistor and its range of possible values, given its color coded percent tolerance.
2. Use an ohmmeter to measure the ohmic value of a resistor.
3. Determine if a resistor is within its coded tolerance.

LAB PREPARATION

Review Section 3.2 of *Introductory Electric Circuits.*

DISCUSSION

The ohmmeter can be used to measure the resistance of a component called a resistor. As will be explained later, the resistor is an element that offers opposition to current. While your instructor will demonstrate the use of your analog or digital multimeter as an ohmmeter, there are several important points you need to remember when measuring resistance:

1. Do not touch the tips of both ohmmeter test leads with your fingers while measuring the resistance of the resistor. Doing so causes the resistance of your skin to interact with the resistance of the resistor, resulting in a false reading.

2. *Never measure the resistance in a circuit with power applied.*
3. Always use the lowest resistance scale that will provide a reading. The lowest resistance scale improves the accuracy of your reading.
4. When using an analog meter, always reset the meter to zero ohms when you change from one range to another.

MATERIALS

1 DMM or VOM

10 assorted resistors (*Note:* No more than three resistors should have any one color in their multiplier band.)

1 protoboard

PROCEDURES

Digital Multimeter

1. Take the ten resistors you have selected and record their color bands in column 1 of Table 2.1.
2. Using the color code, determine the rated value of each resistor. Enter these values in column 2 of Table 2.1.
3. Determine the tolerance of each resistor and enter that percentage in column 3 of Table 2.1.
4. Using the tolerance and rated values, calculate the maximum and minimum values for each resistor. Enter your calculated values in column 4 of Table 2.1.
5. Measure the value of each resistor and enter that value in column 5 of Table 2.1. (Measure the resistance of the resistor by taking the two probes or test leads of the ohmmeter and placing the positive probe tip on one side of the resistor and then taking the negative probe tip and placing it on the other side of

TABLE 2.1 Measured Resistors (DMM)

No.	*Color Bands*	*Nominal Value*	*Percent Tolerance*	*Max./Min. Value*	*Measured Value*	*Percent Error*	*In/out of Tolerance*
1							
2							
3							
4							
5							
6							
7							
8							
9							
10							

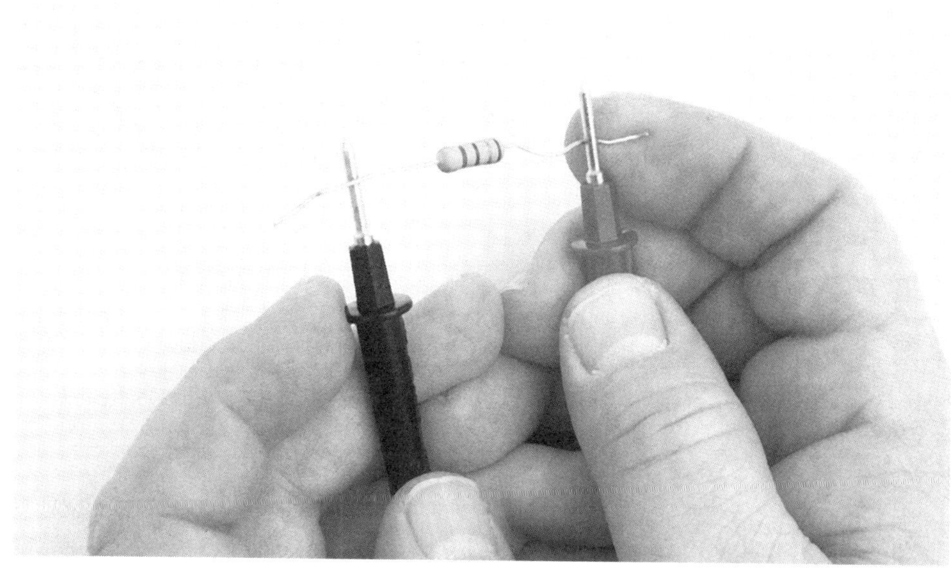

FIGURE 2.1 Measuring resistance.

the resistor. See Figure 2.1. Be sure that your fingers are not touching the metal tips of both probes.)

6. The percent of error between a nominal value and a measured value is found using the formula below.

$$\% \text{ error} = \frac{\text{nominal value} - \text{measured value}}{\text{nominal value}} \times 100$$

Calculate the percent error between the nominal values and the measured values. Enter your calculated values in column 6 of Table 2.1.

7. From column 6, determine which resistors are in tolerance and which resistors (if any) are out of tolerance. Record your findings in column 7 of Table 2.1.

TABLE 2.2 Measured Resistors (VOM)

No.	*Color Bands*	*Nominal Value*	*Percent Tolerance*	*Max./Min. Value*	*Measured Value*	*Percent Error*	*In/out of Tolerance*
1							
2							
3							
4							
5							
6							
7							
8							
9							
10							

Analog Multimeter (Option)

Follow the same procedure that you used for the digital meter. This time you will enter all of your data into Table 2.2. Remember that when you change the range on an analog meter, you must reset the meter to zero. Also, when using an analog meter, you have to look straight into the meter face. Otherwise, you will introduce what is called parallax error. These two concepts will be demonstrated and explained by your instructor.

QUESTIONS

1. What does it mean when the percent of error between the nominal value and the measured value is less than the component's percent of tolerance? ____________

   ____

2. What does it mean when the percent of error between the nominal value and the measured value is greater than the component's percent of tolerance? ____________

   ____

3. A resistor was measured and found to be 1995 Ω. The color code is: red, red, red, silver. Is this resistor in tolerance or out of tolerance? ____________

   ____

4. A resistor was measured and found to be 5200 Ω. The color code is: yellow, violet, red, silver. Is this resistor in tolerance or out of tolerance? ____________

   ____

5. What is the nominal value of a resistor with the following color bands: green, blue, yellow, gold? ____________

   ____

6. If you had to choose between the digital meter or the analog meter to measure your resistors, which one would you want to use? Why? ____________

   ____

   ____

   ____

7. If you used the analog meter, did you notice the parallax effect when you measured your resistors? What was the effect that you saw? ____________

   ____

   ____

   ____

Exercise 3

Potentiometers

OBJECTIVES

After completing this exercise, you should be able to:

1. Determine the rated and adjusted resistance values for a potentiometer.
2. Predict whether the resistance between any two potentiometer leads will increase, decrease, or remain unchanged when you turn the adjustment knob.

LAB PREPARATION

Refer to Section 3.3 of *Introductory Electric Circuits*.

DISCUSSION

Refer to Figure 3.1. The potentiometer is a three-terminal variable resistor whose output resistance is either fixed or varied. The resistance between terminal *A* and terminal *C* normally equals the fixed resistance of some type of resistive material. The resistance of terminal *B* in relationship to terminal *A* and terminal *C* will vary, depending on the position of the wiper-arm of terminal *B*. Figure 3.2 represents the internal structure of a potentiometer and its parts.

MATERIALS

1 10 kΩ linear-taper potentiometer
1 10 kΩ audio-taper potentiometer
1 digital multimeter (DMM)

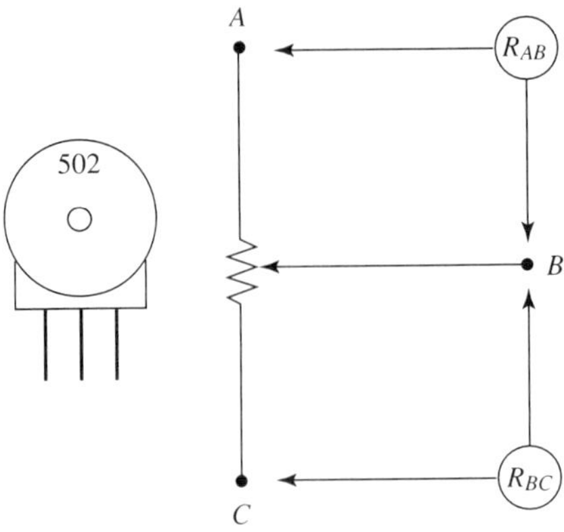

FIGURE 3.1 Pictorial and schematic representation of a potentiometer.

PROCEDURES

Linear Potentiometer

1. Measure and record R_{AC}. R_{AC} = ____________________
2. Using a pencil, make a vertical mark on the control shaft of the potentiometer. Rotate the shaft of the potentiometer fully counterclockwise. Write where the vertical mark on the shaft is on the case of the potentiometer. This position will be position 0.
3. Now rotate the shaft clockwise all the way to the right. Mark as position 1 where the vertical mark on the control shaft points on the potentiometer case. Mark off equal segments to represent one-quarter increments between 0 and 1, as shown in Figure 3.3.
4. With the control shaft fully counterclockwise (which is position 0), measure the values of R_{AC}, R_{AB}, and R_{BC}. Record these values in Table 3.1.
5. Turn the control knob fully clockwise (which is position 1 as marked on the potentiometer case). Measure R_{AC}, R_{AB}, and R_{BC}. Record these values in Table 3.1.
6. Turn the control knob to the half-turn position and mark on the potentiometer case. Measure R_{AC}, R_{AB}, and R_{BC}. Record these values in Table 3.1.

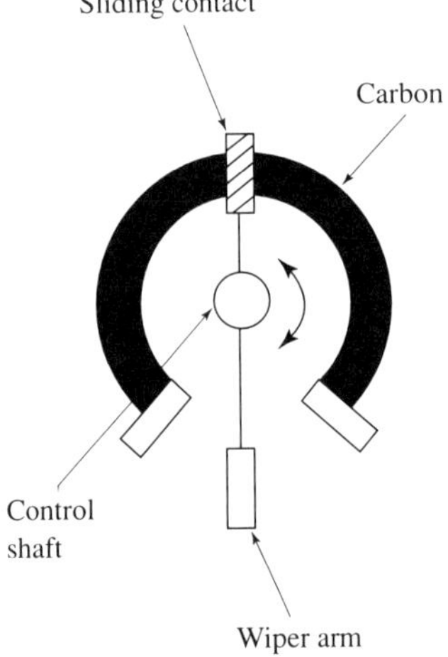

FIGURE 3.2 Internal representation of a potentiometer.

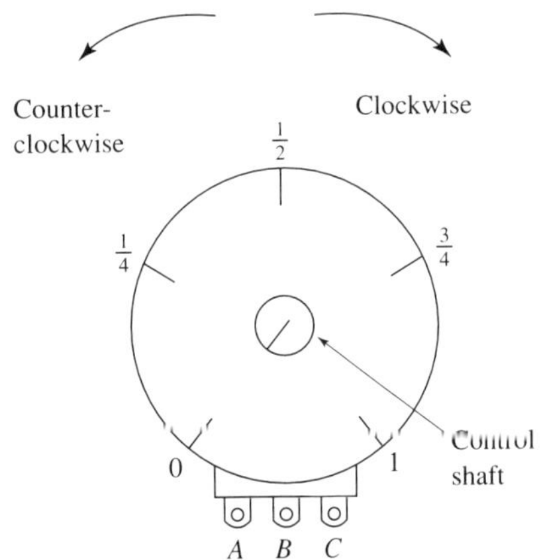

FIGURE 3.3 Positions and electrical connections.

7. Turn the control knob to the quarter-turn position and mark on the potentiometer case. Measure R_{AC}, R_{AB}, R_{BC}. Record these values in Table 3.1.
8. Turn the control knob to the three-quarter-turn position and mark it on the potentiometer case. Measure R_{AC}, R_{AB}, and R_{BC}. Record these values in Table 3.1.

Nonlinear Potentiometer (Option)

1. Measure R_{AC}. R_{AC} = ______________________________
2. Repeat Steps 2 through 8 from the above procedure. Take and record all the measured values in Table 3.2.

TABLE 3.1 Linear Resistance Measurements

Shaft Position	R_{AC}	R_{AB}	R_{BC}
Full left			
Quarter-turn right			
Half-turn right			
Three-quarter-turn right			
Full turn right			

TABLE 3.2 Nonlinear Resistance Measurements

Shaft Position	R_{AC}	R_{AB}	R_{BC}
Full left			
Quarter-turn right			
Half-turn right			
Three-quarter-turn right			
Full turn right			

QUESTIONS

Linear Potentiometer

1. From Table 3.1, pick the measurement of the half-turn position. Does the sum of the measured values of R_{AB} and R_{BC} equal the measured value of R_{AC}?
____
2. Look at the resistance measurements of Table 3.1. What happens to the values of R_{AB} as the values of R_{BC} decrease? Why? ______________________________
____
____
____

3. With the control shaft turned clockwise all the way, describe what would happen to the values of R_{AB} and R_{BC} if you turned the control shaft of the potentiometer counterclockwise for the full rotation. ______

______

4. You have a 100 kΩ potentiometer. You want to set the resistance of the potentiometer (R_{BC}) to 45.5 kΩ with the terminal C at ground potential. How would you set the potentiometer to the resistance of 45.5 kΩ?

______

______

______

Nonlinear Potentiometer (Option)

1. From Table 3.2, pick the measurements of the three-quarter-turn position. Does the sum of the measured values of R_{AB} and R_{BC} equal the measured value of R_{AC}? ______

2. Look at the resistance measurements of Table 3.2. Do the resistance values of R_{AB} increase as the resistance values of R_{BC} decrease? ______

3. Comparing Table 3.1 to Table 3.2, is this increase/decrease relationship a linear or nonlinear relationship? ______

______

4. With the control shaft fully clockwise, what would happen to the resistance values of R_{AB} and R_{BC} if you turned the control shaft counterclockwise through its total rotation? ______

______

______

5. Would these readings be linear or nonlinear? ______

Exercise 4

Circuit Construction

OBJECTIVE

After completing this exercise, you should be able to:

1. Construct any circuit from its schematic.

LAB PREPARATION

Refer to Section 3.7 of *Introductory Electric Circuits.*

DISCUSSION

When you are asked to construct the circuits in the following exercises, you will need to follow some simple concepts and rules:

1. Lab circuits are typically constructed on a structural device called a breadboard (see Figure 4.1). Components, hookup wires, and test leads are inserted into the holes in the breadboard. In Figure 4.1, you can see the individual connecting strips of the vertical structure of the protoboard. You should consider each of those vertical strips as a separate and unique **node:** a point in which you can insert the lead of any component or wire. These nodes are individual points in which components will be connected to each other, as shown in Figure 4.2.
2. Never insert the two lead wires of a component into a single node because you will short out that component.

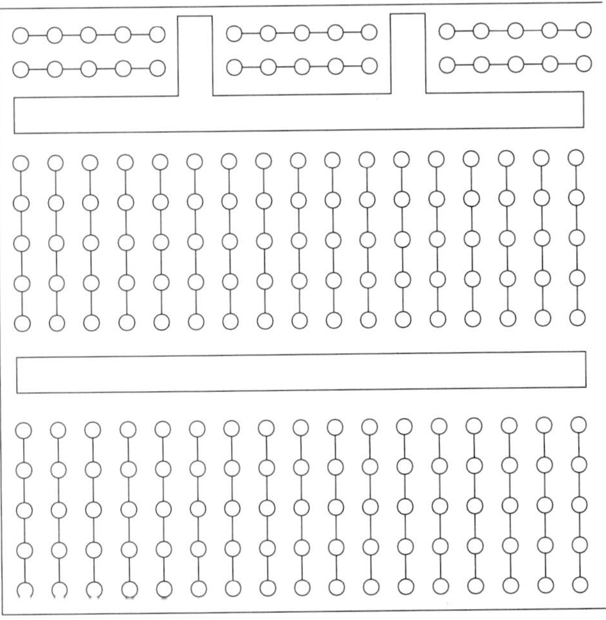

FIGURE 4.1 Typical breadboard internal connections.

3. Whenever you change any of the components in your test circuit, you must turn off the source voltage first.

MATERIALS

1 DMM (digital multimeter)
1 1.2 kΩ, 0.5 W resistor
1 2.2 kΩ, 0.5 W resistor
1 3.3 kΩ, 0.5 W resistor
1 protoboard
1 set of test leads

PROCEDURE

1. Compare the breadboard in Figure 4.1 to the one you are using. Identify the nodes of your breadboard.
2. Measure each of the three resistors and record the measurements in Table 4.1.

TABLE 4.1 Measured Resistance

Nominal Values	*Measured Values*
$R_1 = 1.2\ \text{k}\Omega$	
$R_2 = 2.2\ \text{k}\Omega$	
$R_3 = 3.3\ \text{k}\Omega$	

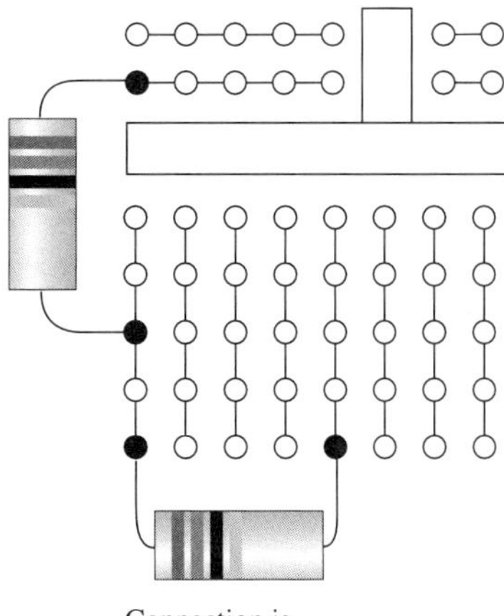

FIGURE 4.2 Component isolation.

Calculate the total resistance of the three resistors and record this value in Table 4.1.

3. In Figure 4.3, identify what we have called a node.
4. Insert one of the leads of a resistor into one of the vertical nodes. This will be node *A*. Insert the lead at the other end of the resistor into an adjacent vertical node, node *B*.
5. Take the second resistor and insert one of the leads into the same vertical node, node *B*, in which you inserted the second lead of the first resistor. Now take the opposite lead of the second resistor and insert it into another unused vertical node, node *C*.
6. Take one of the leads of the third resistor and insert it into the third node, node *C*, where the second resistor was terminated. Then take the lead on the other side of the third resistor and insert it into a fourth node, node *D*. You have now constructed what we call a three-resistor series circuit.
7. Take your DMM and set it up for resistance readings. Take one test lead and clip it to the first resistor at the first node, node *A*. Take the second test lead and clip it to the third resistor lead inserted into the fourth node, node *D*, and read the total resistance from the DMM. Record the resistance reading in Table 4.1.
8. With the DMM, measure the resistance of each resistor from one node to the next. Compare these values to those in Table 4.1. How do the values compare?

____

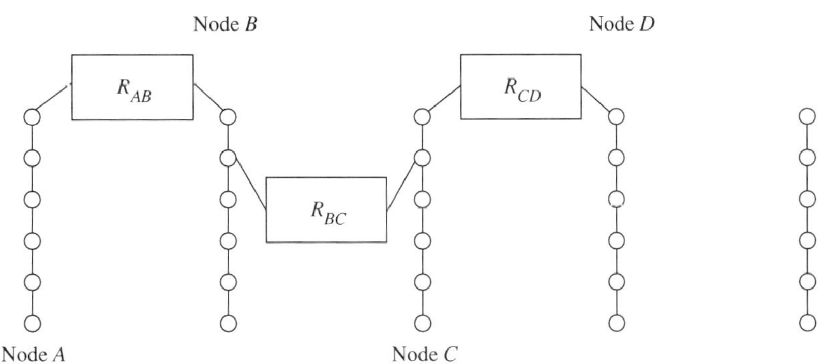

FIGURE 4.3 Test circuit layout.

QUESTIONS

1. Why is it important to identify the separate nodes? ______________________

____

____

2. Why is it important to measure each resistor used in the exercise? ______________

____

____

3. Why isn't it proper procedure to insert the two leads of a resistor into the same node? ____

____

4. Add up the nominal ohmic values of the three resistors in Table 4.1. Is the sum of the nominal values close to the sum of the measured values? ______________

____

____

____

Exercise 5

Measuring Voltage

OBJECTIVES

After completing this exercise, you should be able to:

1. Set up the DMM to take voltage readings.
2. Measure accurately the voltage across any resistor in a circuit.
3. Predict the effects on the voltage across each resistor in a series circuit with a variation of the source voltage.

LAB PREPARATION

Refer to Section 3.8 of *Introductory Electric Circuits*.

DISCUSSION

The source voltage in a basic circuit is distributed among the components in the circuit. For example, each resistor in Figure 5.1 has a measurable voltage across its terminals. Therefore, there are four voltages that can be measured in the circuit: One across each resistor and, of course, the source voltage.

The voltage across any component is measured by connecting a voltmeter across the terminals of that component. For example, to measure the value of V_{AB} in Figure 5.1, the voltmeter leads are connected to nodes A and B. (Remember: Each connection point in the circuit is referred to as a *node*.) When connecting the voltmeter to measure V_{AB}, the "Volts" lead is connected to the positive side of R_{AB}. The "common" lead is connected to the other side. (If you have trouble connecting the voltmeter, review Exercise 1.)

MATERIALS

1 1.2 kΩ resistor
1 2.2 kΩ resistor
1 3.3 kΩ resistor
1 digital multimeter (DMM)
1 analog multimeter (if needed)
1 dc-voltage source
1 protoboard

PROCEDURE

1. Construct the circuit in Figure 5.1. Make sure that the dc power supply is set up properly, as demonstrated in Exercise 1.

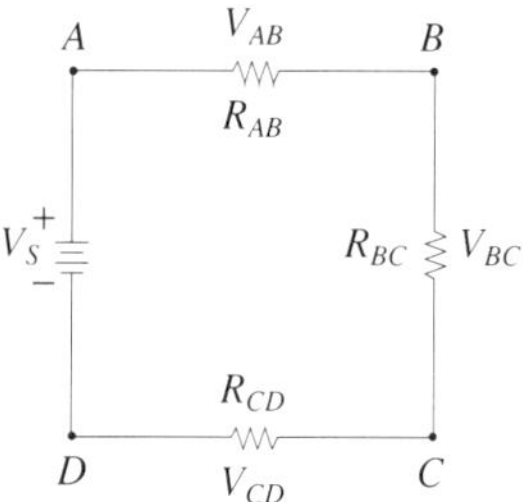

FIGURE 5.1 Nodal circuit construction.

2. Using the function select button (or switch), set the DMM to measure dc volts.
3. Connect the voltmeter (DMM) across the dc power supply inputs to the circuit. (The "volts" lead is connected to node A and the "common" lead is connected to node D.) Turn on the dc power supply and adjust the "+Volts" control to obtain a reading of 10 V on the DMM.
4. Measure the voltage across R_{AB}. Remember: The "volts" lead is connected to the positive side of the component and the "common" lead is connected to the negative side of the component. Record the value of this voltage *(V_{AB})* in the appropriate space in Table 5.1.

TABLE 5.1 Measured Resistors

Nominal Values	*Measured Values*
R_{AB} = 1.2 kΩ	
R_{BC} = 2.2 kΩ	
R_{CD} = 3.3 kΩ	

5. Repeat Step 4 for R_{BC}. Record the value of V_{BC} in the appropriate space in Table 5.1.
6. Repeat Step 4 for R_{CD}. Record the value of V_{CD} in the appropriate space in Table 5.1.
7. Reconnect the DMM to the dc power supply, and increase the value of the source voltage to 15 V.

> Note: If you look at the data in Table 5.2, you should notice that the resistor voltages add up to the value that we used as the magnitude of the source voltage. Changing the source voltage of our circuit will have a direct effect on the voltage drops across the three resistors. You will find this to be true on your next set of voltage readings.

TABLE 5.2 Measured Voltages (Source Is 10 V)

Nodes	*Measured Voltage*
$V_S = V_{AD}$	
V_{AB}	
V_{BC}	
V_{CD}	
V_T	

8. Repeat steps 4, 5, and 6. Record the measured voltages in Table 5.3.

> Note: When you check the sum of the resistor voltages in Table 5.3, you will notice that it is equal to the source voltage.

TABLE 5.3 Measured Voltages (Source Is 15 V)

Nodes	*Measured Voltage*
$V_S = V_{AD}$	
V_{AB}	
V_{BC}	
V_{CD}	
V_T	

QUESTIONS

1. Refer to Table 5.2. Determine the sum of the measured component voltages, and compare it to the measured value of the source voltage. How do these values compare? ____

____

____

2. Refer to Table 5.3. Determine the sum of the measured component voltages, and compare it to the measured value of the source voltage. How do these values compare? ____

____

____

3. Compare the readings in Table 5.2 to those in Table 5.3. What happens to the voltage across each resistor when the source voltage is increased? ____________

____

Based on your answer, what do you think would happen to the resistor voltages if we were to *decrease* the value of the source voltage? ____________________

____

4. Briefly describe, in your own words, the procedure for measuring the voltage across any resistor in a circuit like the one in Figure 5.1. __________________

____

____

____

Exercise 6

Measuring Current

OBJECTIVES

After completing this exercise, you should be able to:

1. Set up a DMM to measure current.
2. Measure accurately the current at any given node or test point in a circuit.
3. Predict the effects of each of the following on the total current in a series circuit:
   a. An increase or decrease in supply voltage.
   b. An increase or decrease in total circuit resistance.
4. State the relationship between the currents through any node, test point, or element in a series circuit.

LAB PREPARATION

Refer to Section 3.8 of *Introductory Electric Circuits.*

DISCUSSION

The circuit in Figure 6.1 is referred to as a *series circuit.* A series circuit is one that provides only one path for current. For example, if you trace the circuit from one side of the source to the other, you'll see that:

1. There is only one path between the terminals of the source.
2. The single path contains all the elements (resistors) in the circuit.

Series circuit characteristics are covered in Chapter 5 of *Introductory Electric Circuits*. However, for our purposes, we need to establish one important characteristic at this time: *Current maintains a constant value throughout a series circuit*. In other words, the current measured at any point in the circuit will always equal the current measured at any other point. This relationship is demonstrated in this exercise.

To measure dc current, a DMM (or VOM) is set to function as a *dc milliammeter*. The level of current at any point in a dc circuit is measured by opening the circuit at the given node (or test point) and inserting the ammeter in series with the circuit elements. This procedure is demonstrated in this exercise.

Measuring Current

When used as an ammeter, most DMMs need to be set up first. Some DMMs have a separate mA input jack. The black test lead will still be connected to the common or low input jack. However, the red test lead should be connected to the mA input jack of the DMM. Once the test leads are connected correctly to the DMM, make the following adjustments:

1. Select the dc mA function mode on the meter.
2. Select the highest mA range on the selector switch.

MATERIALS

1 1.2 kΩ, 0.5 W resistor
1 2.2 kΩ, 0.5 W resistor
1 3.3 kΩ, 0.5 W resistor
1 10 kΩ, 0.5 W resistor
1 dc power supply
1 DMM
1 protoboard

TABLE 6.1 Measured Resistor Values

Nominal Values	*Measured Values*
$R_1 = 1.2$ kΩ	
$R_2 = 2.2$ kΩ	
$R_3 = 3.3$ kΩ	
$R_4 = 10$ kΩ	

PROCEDURE

1. Construct the circuit shown in Figure 6.1.
2. Set the DMM to read dc volts. With the circuit connected as shown, adjust the source voltage to 10 V. Now turn off the source voltage and do not touch the voltage adjust knob.

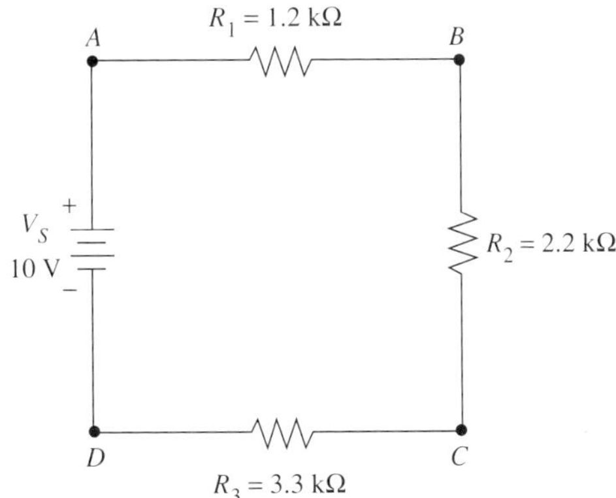

FIGURE 6.1 Current test circuit.

3. Set the DMM to read dc mA.
4. Insert the ammeter into the circuit at node *A*. To do this, open the circuit by taking out the resistor lead inserted in node *A*. Place the ammeter in the circuit with the red positive test lead connected to the positive terminal of the battery supply. Clip the black test lead to the end of the resistor lead you just took out of node *A*. Now turn on the source voltage (which has already been set to 10 V). Adjust the range scale down on the ammeter until you get the best possible reading. Record this reading in Table 6.2. Again, turn off the supply source, remove the ammeter from the circuit, and restore the circuit to its original condition.
5. Do the same thing for node *B*. At this time the ammeter should be set to its highest range. Open the circuit at node *B*, and insert the ammeter between resistors R_1 and R_2. The red lead is clipped to the loose end of R_1 and the black lead is clipped to the loose end of R_2. Turn on the source voltage and check to see if you are on the best range to get the best current reading. Record that reading in Table 6.2. Turn off the voltage source and restore the circuit to its original condition.
6. Insert the ammeter at node *C*. Set the ammeter range to its highest range. Open the circuit at node *C*, and insert the ammeter between R_2 and R_3. Turn on the voltage source and select the right range. Record the reading in Table 6.2. Turn off the source voltage and restore the circuit.
7. The last reading will be taken at node *D*. Open the circuit at node *D*. Place the red test lead of the ammeter to the loose end of R_3 and the negative test lead of the ammeter to the negative terminal of the voltage source. Turn on the voltage source and select the right current range. Record the reading in Table 6.2.

TABLE 6.2 Currents with Source Voltage at 10 V

Node	*Current at Node*
A	
B	
C	
D	

TABLE 6.3 Currents with Source Voltage at 15 V

Node	*Current at Node*
A	
B	
C	
D	

8. Restore the circuit to its original setup. Using the DMM as a voltmeter, reset the source voltage to 15 V. Repeat Steps 3 through 7 for the new voltage level. Record these currents in Table 6.3.
9. Turn off the voltage source and replace resistor R_2 with a 10 kΩ resistor.
10. Turn on the voltage source and adjust the voltage for 15 V.
11. Repeat Steps 3 through 7 for the new resistance. Record the new current reading in Table 6.4.

TABLE 6.4 Currents with R_2 at 10 kΩ

Node	*Current at Node*
A	
B	
C	
D	

QUESTIONS

1. From Table 6.2, compare the current readings at Nodes *A*, *B*, *C*, and *D*. Based on your readings, what can you say about the current in a series circuit?

____

____

____

2. Compare the current readings from Table 6.2 and Table 6.3. Based on these readings, what is the relationship between source voltage and circuit current?

____

____

____

3. Compare the current readings from Table 6.3 and Table 6.4. Based on these readings, what is the relationship between the current in the circuit and the resistance of the circuit? ______________________________

____

____

____

4. In your own words, discuss what you have learned from this exercise.

Exercise 7

Ohm's Law

OBJECTIVE

After completing this exercise, you should be able to:

1. Verify, through circuit measurements, the relationship among voltage, current, and resistance.

LAB PREPARATION

Review Section 4.1 of *Introductory Electric Circuits*.

DISCUSSION

In this exercise, we will take a look at the relationship among current, resistance, and voltage. Ohm's law states that current is directly proportional to voltage and inversely proportional to resistance, and is expressed as follows:

$$I = \frac{V}{R}$$

where

I = current

V = voltage

R = resistance

You will take several circuit current measurements for each of the individual resistors in Table 7.1 to verify this relationship.

MATERIALS

1 dc voltage source
1 DMM
1 1 kΩ resistor
1 2 kΩ resistor
1 10 kΩ resistor
1 protoboard
1 set of test leads

TABLE 7.1 Resistor Measurements

Resistor	*Measured*
1 kΩ	
2 kΩ	
10 kΩ	

PROCEDURE

1. Measure all the resistors and record their values in Table 7.1. Use these values for your circuit calculations.
2. Construct the circuit shown in Figure 7.1 using the first resistor in Table 7.1. Using the DMM, set V_S to 10 V. Now set the DMM to measure current.
3. Predict the value of I_S in the circuit using Ohm's law. Enter the predicted value of current in the appropriate space in Table 7.2.

> Note: Always use the measured values of the resistors to calculate (predict) the value of circuit current.

4. Measure the current in the circuit at either node. Record your measured value in the appropriate space in Table 7.2.
5. For each combination of V and R shown in Table 7.2, predict the individual currents. Record these values in Table 7.2.

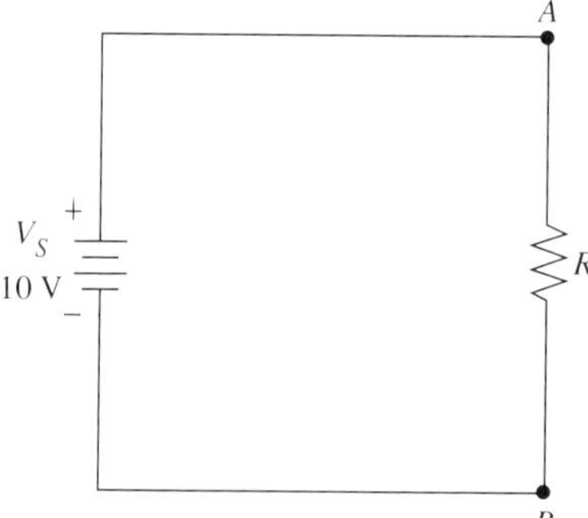

FIGURE 7.1 Test circuit for Ohm's law.

TABLE 7.2 Current Measurements

V	*R*	*I (Predicted)*	*I (Measured)*	*Percent of Error*
10 V	1 kΩ			
10 V	2 kΩ			
10 V	10 kΩ			
12 V	10 kΩ			
15 V	10 kΩ			

6. Using the same circuit configuration, measure the current for each of the combinations of voltage and resistance left in Table 7.2. Record these current readings in Table 7.2.

QUESTIONS/PROBLEMS

1. What is the relationship between current and a fixed resistance when the voltage is increased? Explain your answer using values recorded in Table 7.2.

2. What is the relationship between fixed voltage and current when resistance is increased? Explain your answer using values recorded in Table 7.2.

3. Calculate the percent of error between your predicted and measured current values. Record the percent of error in the appropriate spaces in Table 7.2.

4. You predicted the current for each set of voltage and resistance values in Table 7.2 using the measured values of the resistors. Try using the nominal value of two of the resistors and discuss the different results.

1 kΩ (current value) ____________________

____

2 kΩ (current value) ____________________

____

Is there a difference in the predicted currents as opposed to the measured currents? If there is a difference, explain why. ____________________

____

____

5. In your own words, discuss what you observed in this exercise.

Exercise 8

Series Circuit Characteristics

OBJECTIVES

After completing this exercise, you should be able to:

1. Calculate the total resistance of a series circuit.
2. Determine the component voltages of a series circuit.
3. Determine the power relationship for a series circuit.
4. State the characteristics that distinguish a series circuit from other types of circuits.

LAB PREPARATION

Review Section 5.1 of *Introductory Electric Circuits*.

DISCUSSION

In this exercise you will determine the characteristics of a series circuit. In Exercise 6, you discovered that the current through a series circuit is the same at any point of the circuit. In this exercise, you will explore the other characteristics of a series circuit. The following equations will help you.

$$I_S = I_1 = I_2 = I_3 = \ldots = I_N \qquad R_T = R_1 + R_2 + R_3 + \ldots + R_N$$

$$V_S = V_1 + V_2 + V_3 + \ldots + V_N \qquad P_S = P_1 + P_2 + P_3 + \ldots + P_N$$

X_S = total unit value

X_N = the highest subscriptive numbered unit value

MATERIALS

1 dc power supply
1 DMM
1 protoboard
1 1 kΩ, 0.5 W resistor
1 3.3 kΩ, 0.5 W resistor
1 2.2 kΩ, 0.5 W resistor
1 set of test leads

PROCEDURE

1. In Table 8.1, record the nominal values of R_1, R_2, and R_3 from the test circuit in Figure 8.1.

TABLE 8.1 Resistance Values

Nominal Resistor Value	*Measured Resistor Value*
$R_1 = 2.2\ k\Omega$	
$R_2 = 3.3\ k\Omega$	
$R_3 = 1\ k\Omega$	
$R_T = 6.5\ k\Omega$	

2. Measure the ohmic values of the three resistors. Record these values in Table 8.1.
3. Construct the circuit in Figure 8.1 without the power supply connected.
4. Using the measured resistor values, calculate the total circuit resistance. Record this value in Table 8.1.
5. With the power disconnected, measure the total resistance, which would be from node *A* to node *D*. Record the total measured resistance in Table 8.1.

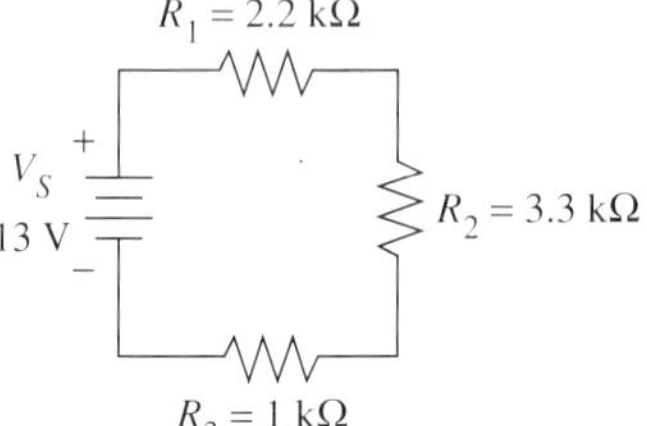

FIGURE 8.1 Series test circuit.

TABLE 8.2 Series Circuit Current

Calculated I_S =
Measured I_S =

TABLE 8.3 Series Circuit Voltage

Calculated Voltages	*Measured Voltages*
V_1 =	V_1 =
V_2 =	V_2 =
V_3 =	V_3 =
V_T =	V_T =

TABLE 8.4 Series Circuit Power Dissipation

Predicted Power
P_1 =
P_2 =
P_3 =
P_T =

6. Using Ohm's law, predict the value of I_S. Be sure to use the measured values of V_S and R_T. Record the predicted current in Table 8.2.
7. Connect the power supply and, with the DMM, set the voltage to 13 V.
8. Using your DMM as a current meter, measure the source circuit current at any point of the test circuit. Record this value in Table 8.2.
9. Using your measured resistor values and the measured value of I_S, calculate the voltage developed across each of the three resistors. Record these voltage values in Table 8.3.
10. Measure the voltages across the three resistors and record these voltages in Table 8.3.
11. Using the measured voltage values of Step 10 and the measured current values of I_S, calculate the measured power dissipated by each resistor. Record these calculated values in Table 8.4.
12. Using Ohm's law, calculate the total power developed by the source and delivered to the circuit. Record this value in Table 8.4.

QUESTIONS

1. Does the sum of the individual measured resistors equal the total resistance as measured in Step 5? ____________________

2. In your own words, state the relationship of total resistance in a series circuit in terms of the individual resistances. ______

3. Does the sum of the voltage drops in Step 10 equal the source voltage of 13 V? ______

4. In your own words, state the series circuit relationship of the individual voltage drops across the individual resistors and the source voltage. ______

5. Is the power dissipated by the individual resistors of Step 11 equal to the power provided by the source voltage as calculated in Step 12? ______

6. In your own words, state the series circuit relationship between power dissipated and power provided by the source. ______

7. In your own words, state the series circuit relationships among the individual and total values of resistance, current, voltage, and power.

   a. Resistance: ______

   b. Current: ______

   c. Voltage: ______

   d. Power: ______

Exercise 9

Series Circuits: Kirchhoff's Voltage Law

OBJECTIVE

After completing this exercise, you should be able to:

1. Apply Kirchhoff's voltage law to a series circuit configuration.

LAB PREPARATION

Review Section 5.2 of *Introductory Electric Circuits.*

DISCUSSION

In this exercise, you will prove Kirchhoff's voltage law. This law states that the sum of all component voltage drops in a series circuit is equal to the source voltage.

MATERIALS

1 dc power supply
1 DMM
1 470 Ω, 0.5 W resistor
1 3.3 kΩ, 0.5 W resistor
1 10 kΩ, 0.5 W resistor

1 protoboard
1 set of test leads

PROCEDURE

1. Measure all resistor values and record the measurements in Table 9.1.
2. Make sure that the test leads of your DMM are red (positive) and black (negative). The red lead must be plugged into the positive input jack of your meter. The black test lead must be plugged into the negative, or ground, input jack of your meter.

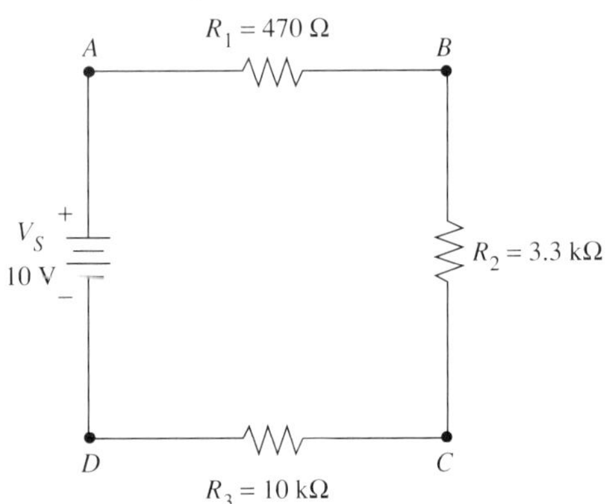

FIGURE 9.1 Kirchhoff's voltage law test circuit.

3. Construct the test circuit in Figure 9.1. Using the DMM as a voltmeter, adjust the supply voltage to 10 V.
4. Using the DMM as a current meter, pick a point in the circuit and measure the source current. I_S = ______________________
5. Calculate the individual voltage drops using the measured value of I_S and the measured resistances. Record these calculated voltage drops in Table 9.2.
6. Total the calculated voltage drops and record this value as V_S in Table 9.2.
7. Now using the same test circuit, measure the voltage across R_1. Place the red lead on the left side of R_1, as in Figure 9.1, and the black lead on the right side of R_1. Record this voltage drop in Table 9.3. Indicate polarity.

> Note: The readout on the DMM will indicate a positive voltage reading. If you get a negative reading, you have reversed the test leads.

TABLE 9.1 Measured Resistors

Resistors	*Measured Values*
$R_1 = 470\ \Omega$	
$R_2 = 3.3\ k\Omega$	
$R_3 = 10\ k\Omega$	
$R_T = 13.77\ k\Omega$	

TABLE 9.2 Circuit Voltage Drops

Calculated Voltages	*Measured Voltages*
V_1 =	V_1 =
V_2 =	V_2 =
V_3 =	V_3 =
V_S =	V_S =

TABLE 9.3 Positive Voltage Polarity

Positive Voltage Readings
V_1 =
V_2 =
V_3 =
V_S =

TABLE 9.4 Negative Voltage Polarity

Negative Voltage Readings
V_1 =
V_2 =
V_3 =
V_S =

8. Repeat this same procedure for R_2 and R_3. Make sure that you have your test leads positioned at the correct points, across each resistor in the circuit. Record these voltage readings in Table 9.3. Indicate polarity.
9. Measure the power supply voltage with the red lead at the top of the source supply and the black lead at the bottom of the source supply. Record this reading as V_S in Table 9.3. Indicate polarity.
10. Repeat Steps 7 through 9, but reverse the test lead placement for each reading. Record these readings in Table 9.4. Indicate polarity.

QUESTIONS

1. Is the sum of the measured voltage drop readings in Table 9.2 approximately equal to the source voltage setting? ______________________
2. Is the sum of the predicted component voltage drops in Table 9.2 approximately equal to the source voltage? ______________________

3. Are the measured voltage drops in Table 9.3 approximately equal to the measured source voltage? ____________

4. Are the measured voltage drops in Table 9.4 approximately equal to the measured source voltage? ____________

5. Are the polarities of the voltage drops in Table 9.3 opposing or aiding the polarity of the voltage source? ____________

____

6. Explain the importance of knowing the polarity of the test probes.

____

____

7. What is the significance of Step 4 with regard to Kirchhoff's voltage law?

____

____

____

8. Write two equations, one each for Table 9.3 and Table 9.4, that will satisfy Kirchhoff's voltage law.

   a. ____________

   b. ____________

9. Explain the significance of being aware of the polarity of the voltage drops across components in an electrical circuit. ____________

____

____

Exercise 10

Series Circuits: Voltage Divider Rule

OBJECTIVE

After completing this exercise, you should be able to:

1. Apply the voltage divider rule to a series circuit configuration.

LAB PREPARATION

Review Section 5.2 of *Introductory Electric Circuits.*

DISCUSSION

In previous exercises, you have discovered that:

1. The total resistance of a series circuit is the sum of the individual resistor values.
2. The current is the same at any point in the circuit.
3. The sum of the voltage drops in a series circuit is equal to the source voltage.

In this exercise, you will calculate the voltage drop across each resistor before you turn on the source voltage. All that you need to know to find this voltage drop is the total resistance of the circuit, the magnitude of the source voltage, and the value of the resistance for which you want to find the voltage drop. The equations that you will need are given below.

$$V_N = \frac{R_N \times V_S}{R_T} \qquad \text{Percent difference} = \left| \frac{V_{\text{CALC}} - V_{\text{MEAS}}}{V_{\text{MEAS}}} \right| \times 100$$

MATERIALS

1 dc power supply
1 DMM
1 2 kΩ, 0.5 W resistor
1 4.7 kΩ, 0.5 W resistor
1 3.3 kΩ, 0.5 W resistor
1 protoboard
1 set of test leads

PROCEDURE

TABLE 10.1 Measured Resistor Values

Resistors	*Measured*
1.2 kΩ	
2.2 kΩ	
3.3 kΩ	

1. Measure all three resistors and record their values in Table 10.1.
2. Construct the test circuit in Figure 10.1.
3. Turn on the voltage source. Using the DMM as a voltmeter, set the source voltage at 10 V.
4. Using the voltage divider rule, predict the voltage drop across R_1. With the DMM, measure the actual voltage drop across R_1. Record both values in Table 10.2.
5. Using the voltage divider rule, predict the voltage across R_2. Measure the actual voltage drop across R_2. Record both voltage values in Table 10.2.
6. Predict the voltage drop across R_3 of the test circuit. With the DMM, measure the voltage across the resistor R_3. Record these values in Table 10.2.

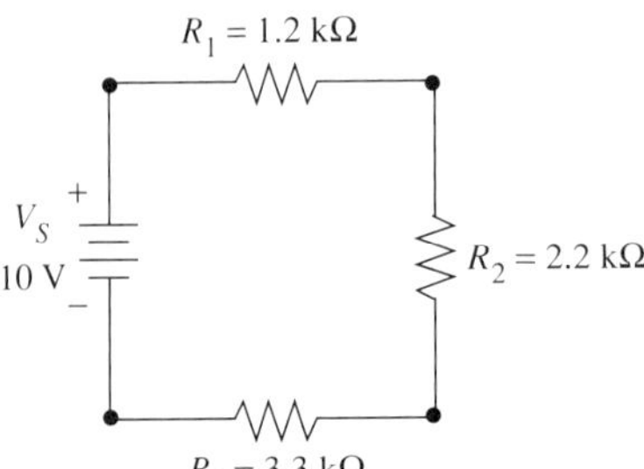

FIGURE 10.1 Voltage divider rule test circuit.

TABLE 10.2 Voltage Comparisons

Resistors	*Calculated Voltage*	*Measured Voltage*	*Percent of Error*
1.2 kΩ			
2.2 kΩ			
3.3 kΩ			
Totals			

TABLE 10.3 Voltage Comparisons

Resistors	*Calculated Voltage*	*Measured Voltage*	*Percent of Error*
R_1 and R_2			
R_2 and R_3			

7. Using the percent difference equation, calculate the percent of error between the calculated and measured voltage values in Table 10.2. Record the percent of error for each set of voltage readings for V_1, V_2, and V_3 in Table 10.2.
8. Using the voltage divider rule, predict the voltage drop across the combination of R_2 and R_3 ($R_2 + R_3 = R_N$). Now measure the voltage drop across the combination of R_2 and R_3 and record this voltage in Table 10.3.
9. Predict the voltage across resistors R_1 and R_2 as if they were one resistor. Now measure the voltage drop across R_1 and R_2. Remember to use always your measured resistor values in your predictions. Record this voltage value in Table 10.3.
10. Calculate the percent of error for the two sets of voltage readings and record this calculation in Table 10.3.

QUESTIONS

1. Are all the predicted voltages of Table 10.2 approximately equal to the measured voltage values? What is the largest percent of error? ____________________
2. In Table 10.3, is the combined predicted voltage drop across R_2 and R_3 approximately equal to the measured voltage? What is the percent of error? __________
3. In Table 10.3, is the predicted voltage drop across R_1 and R_2 approximately equal to the measured voltage value of R_1 and R_2? What is the percent of error? ____
4. In your own words, state the advantage of using the voltage divider rule when analyzing the voltage drop in a series circuit configuration. ______________

____

____

____

____

Exercise 11

Measuring Power

OBJECTIVES

After completing this exercise, you should be able to:

1. Determine the power dissipated by a resistor.
2. Determine the proper power ratings of the resistors used in the exercise.
3. Determine the total power delivered and dissipated by a circuit.

LAB PREPARATION

Review Section 5.3 of *Introductory Electric Circuits.*

DISCUSSION

In this exercise you will learn how to determine the power dissipation of the total circuit and of the individual circuit resistors. You will need to know if the resistors that you use in your circuits have a large enough power rating and will not overheat and burn. Below are the equations useful for this exercise.

$P_S = I_T \times V_S$ $\qquad$ $P_N = I_T \times V_N$

P_S = total power $\qquad$ V_S = source voltage $\qquad$ I_T = total current

P_N and V_N are used to find the power dissipated by an individual resistor.

MATERIALS

1 dc power supply
1 DMM
1 330 Ω, 0.5 W resistor
1 330 Ω, 0.25 W resistor
1 220 Ω, 0.5 W resistor
1 100 Ω, 0.5 W resistor
1 set of test leads

PROCEDURE

TABLE 11.1 Resistor Measurements

Nominal Values	*Measured Values*
$R_1 = 330\ \Omega$	
$R_2 = 220\ \Omega$	
$R_3 = 100\ \Omega$	
$R_4 = 330\ \Omega$, 0.25 W	

1. Measure the resistors for this exercise and record the values in Table 11.1.
2. Use the 0.5 W resistors to construct the circuit shown in Figure 11.1.
3. Using the DMM as a voltmeter, set the source voltage for 13 V.
4. At any one of the nodes (*A*, *B*, *C*, or *D*), insert the DMM as a dc current meter, set on the 1A range. Apply power to the circuit and decrease the current range until you get a good current reading. Touch the 330 Ω resistor and determine if it is very warm. Record the current reading in Table 11.1. Turn off the source voltage. Since this is a series circuit, what is the value of current through all the resistors? Explain. ____

   ____
5. Now using the same circuit, remove the 330 Ω, 0.5 W resistor and replace it with the 330 Ω, 0.25 W resistor. Repeat Steps 3 and 4. Don't forget to record the

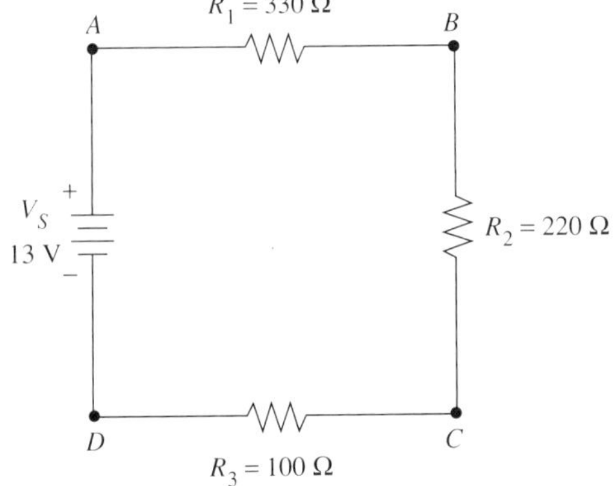

FIGURE 11.1 Power test circuit.

TABLE 11.2 Power Measurements

Resistors	*Measured Voltage*	*Measured Current*	*Power Dissipated*
R_1 0.5 W			
R_2 0.5 W			
R_3 0.5 W			
R_4 0.25 W			
			P_S =

resistance of the new resistor and the current for this circuit in Table 11.2. Touch the 330 Ω resistor and determine if it is very warm.

QUESTIONS

1. Using the information from Table 11.2, what was the power dissipation for each resistor?

   330 Ω, 0.5 W ______

   220 Ω, 0.5 W ______

   100 Ω, 0.5 W ______

   330 Ω, 0.25 W ______

2. What was the total power supplied by the source? ______
3. Is the sum of the individual power dissipations of the resistors equal to the power delivered to the circuit by the source? ______
4. What conclusions can you draw from your answers to the first three questions?

   ______

   ______

   ______

5. Was the 330 Ω, 0.25 W resistor warmer than the 330 Ω, 0.5 W resistor? Why?

   ______

   ______

   ______

6. What is the relationship between the physical size of a resistor and its power dissipation rating? ______

   ______

   ______

   ______

Exercise 12

Series Circuits: Troubleshooting Faults

OBJECTIVES

After completing this exercise, you should be able to:

1. Recognize the effect of an open circuit in a series circuit.
2. Calculate and measure the voltage across an open circuit.
3. Recognize the effect of a shorted resistor in a series circuit.
4. Calculate and measure the current through a shorted component.

LAB PREPARATION

Review Section 5.3 of *Introductory Electric Circuits*.

DISCUSSION

One of the skills that you will have to develop is troubleshooting faults in an electrical circuit. You need to learn how to recognize open circuits that disrupt current and short circuits that increase current. You also need to know what effect these conditions have on the overall circuit.

MATERIALS

1 dc power supply
1 DMM

1 120 Ω, 0.5 W resistor
1 220 Ω, 0.5 W resistor
1 180 Ω, 0.5 W resistor
1 protoboard
1 set of test leads

PROCEDURE

TABLE 12.1 Measured Resistance

Resistors	*Measured Resistors*
120 Ω	
220 Ω	
180 Ω	

1. Measure all resistors and record their measured values in Table 12.1.
2. Construct the circuit in Figure 12.1.
3. With your DMM, set the source voltage, V_S, at 12 V. Record this value in Table 12.2.
4. Using the DMM set up for current, measure the normal circuit current, I_S. With the DMM set up for voltage, measure the normal voltage drops across each resistor, R_1, R_2, and R_3. Record these three voltages in Table 12.2.
5. To simulate an open circuit, remove R_2 (between nodes B and C).
6. Measure the current, I_S. *Note: Do not place the ammeter between nodes B and C,* as this would restore the circuit current path. Measure the voltage drops across R_1 and R_2. Record these two values in Table 12.2.
7. Measure the voltage drop across nodes B and C (where R_2 had been). Record this value as V_2 in Table 12.2.
8. Turn off the voltage source and restore the circuit to its original configuration.
9. Turn on the voltage source and make sure that the source voltage is still set to 12 V.
10. Measure the source current, I_S. Measure the voltage drops V_1, V_2, and V_3. Record these four values in Table 12.3.
11. Take a jumper wire and short out R_2. (Insert one end of the jumper wire at node B and the other end of the jumper wire at node C.)
12. Measure the source current, I_S, between the positive terminal of the power supply and node A. Record this value in Table 12.3.
13. Measure the voltage across each resistor and record the readings in Table 12.3.

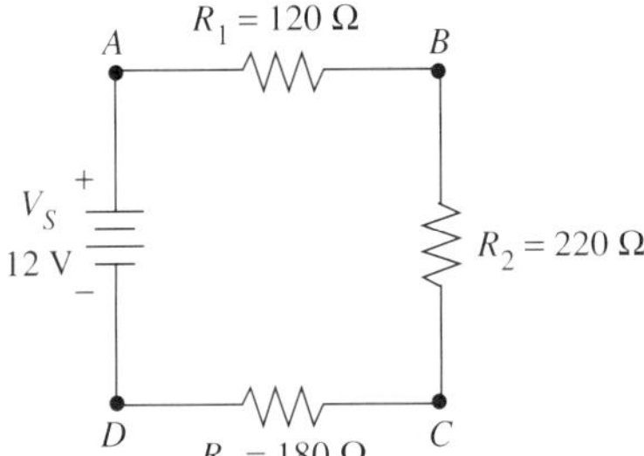

FIGURE 12.1 Troubleshooting test circuit.

TABLE 12.2 Open Circuit Measurements

Circuit Conditions	
Normal	*Open*
$I_S =$	$I_S =$
$V_1 =$	$V_1 =$
$V_2 =$	$V_2 =$
$V_3 =$	$V_3 =$

TABLE 12.3 Short Circuit Measurements

Circuit Conditions	
Normal	*Shorted*
$I_S =$	$I_S =$
$V_1 =$	$V_1 =$
$V_2 =$	$V_2 =$
$V_3 =$	$V_3 =$

QUESTIONS

Open Circuit

1. With the circuit open, what was the measured current? ____________________
2. With the circuit open, what was the voltage drop across R_1? $V_1 =$ __________
3. With the circuit open, what was the voltage drop across R_2? $V_2 =$ __________
4. With the circuit open, what was the voltage drop across R_3? $V_3 =$ __________
5. With the circuit open, why is the source current, I_S, the value that it is?

____

____

____

6. Explain the open-circuit values of V_1 and V_3 in Table 12.2.

____

____

____

7. Explain the open-circuit value of V_2 in Table 12.2. ____________________

____

____

8. In your own words, explain the series circuit action when there is an open component or components. ______________________________

____

____

____

Short Circuit

9. From Table 12.3, is there a difference in the two currents? Why? __________

____

____

10. From Table 12.3, is there a difference in the voltage readings of V_1? __________

11. From Table 12.3, is there a difference in the voltage readings of V_3? __________

____

12. In your own words, explain why there is a difference in the voltage readings of Questions 10 and 11. ______________________________

____

____

____

13. List five circuit conditions that you should look for when you troubleshoot a circuit for either a short or an open circuit.

a. ______________________________

____

b. ______________________________

____

c. ______________________________

____

d. ______________________________

____

e. ______________________________

____

Exercise 13

dc Source Resistance

OBJECTIVE

After completing this exercise, you should be able to:

1. Measure the internal resistance of a dc power supply.

LAB PREPARATION

Review Section 5.4 of *Introductory Electric Circuits.*

DISCUSSION

Every voltage source has some amount of internal resistance. This resistance is in series with any circuit or network connected to the voltage source. Figure 13.1 shows the simple equivalent circuit representing this concept. Since the source resistance is in series with the load, the two resistances form a voltage divider relationship. Thus, the actual load voltage is found as:

$$V_L = V_S \times \frac{R_L}{R_T}$$

where

R_T = the sum of the load resistance (R_L) and the source resistance (R_S)

In this exercise, we will use a decade box to measure the internal resistance of a voltage source.

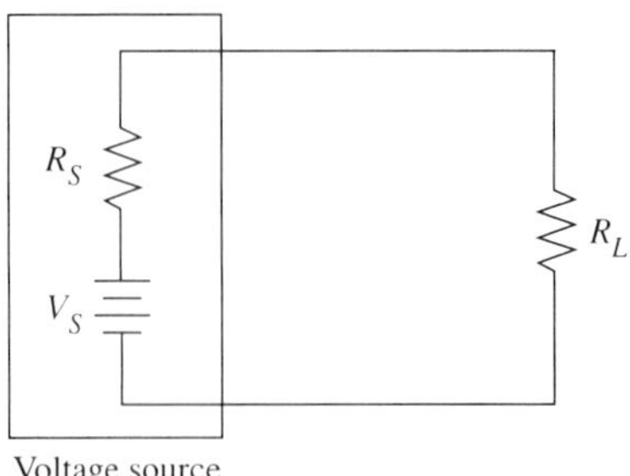

FIGURE 13.1 Simple source voltage and resistance.

MATERIALS

1 decade resistance box
1 dc power supply
1 DMM
2 sets of test leads

PROCEDURE

1. Construct the circuit of Figure 13.2. Make sure that:
   a. The current limiting control on your power supply is set to its minimum setting (if applicable).
   b. Your decade resistance box is set to 2 kΩ.
2. Set the output of the dc power supply to 4 V.
3. Connect the DMM to read the voltage across the decade box.
4. Slowly decrease the setting of the decade box until the voltage across the decade box is equal to 2 V (one-half of the original power supply setting).
5. Record the decade box resistance below. This value is approximately equal to the source resistance (R_S).

   R_S = ___________
6. If your power supply has one, slowly turn the current limiting control up. Describe what happens to the voltage across the decade box. ___________

   ____

   ____

FIGURE 13.2 Source resistance test circuit.

QUESTIONS

1. Using the equation given in the introduction, explain why the ideal voltage source would have no internal resistance. ______________________________

____

____

2. Why did we assume that the source resistance equals the reading from the decade box in Step 5? ______________________________

____

____

____

3. What happened to the internal resistance of your power supply when the current limiting control changed? (*Hint:* Consider your observations in terms of the above equation.) ______________________________

____

____

Exercise 14

Aiding and Opposing dc Voltage Sources

OBJECTIVES

After completing this exercise, you should be able to:

1. Calculate and measure the effect of series-aiding dc voltage sources in the same circuit.
2. Calculate and measure the effect of series-opposing dc voltage sources in the same circuit.

LAB PREPARATION

Review Section 5.4 of *Introductory Electric Circuits*.

DISCUSSION

Electronic circuits can have more than one voltage source. When two voltage sources are connected in series, the total voltage is equal to the sum of, or difference between, the two sources. When connected as *series-aiding* sources, the total voltage equals the sum of the individual source voltages. When connected as *series-opposing* sources, the total voltage equals the difference between the individual source voltages.

In this exercise, you will make some circuit calculations and measurements involving series-aiding and series-opposing sources. You will also be given the opportunity to develop several voltage sources by combining batteries and your dc power supply.

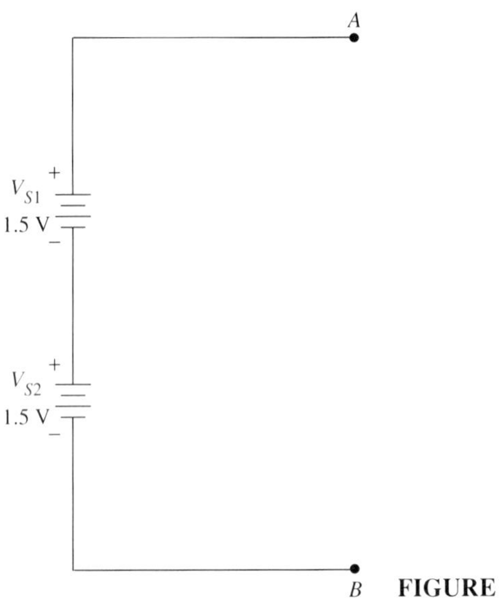

FIGURE 14.1 Series-aiding circuit.

MATERIALS

1 dc power supply
1 DMM
2 1.5 V D-cell batteries
2 D-cell battery holders
1 9 V battery

PROCEDURES

Series-Aiding dc Sources

1. Construct the circuit of Figure 14.1 by hooking up the two 1.5 V batteries as shown. Calculate the total voltage across nodes *A* and *B* and record this value in Table 14.1.

TABLE 14.1 Series-Aiding Measurements

Calculated Voltage	*Measured Voltage*	*Percent of Error*

2. Take your multimeter, set it up to read dc voltage, and measure the voltage across nodes *A* and *B*. Record this voltage in Table 14.1 next to the calculated voltage.
3. Compare the measured voltage to the calculated voltage, and determine the percent of error. ______________________________
4. Comment on the amount of error. ______________________________

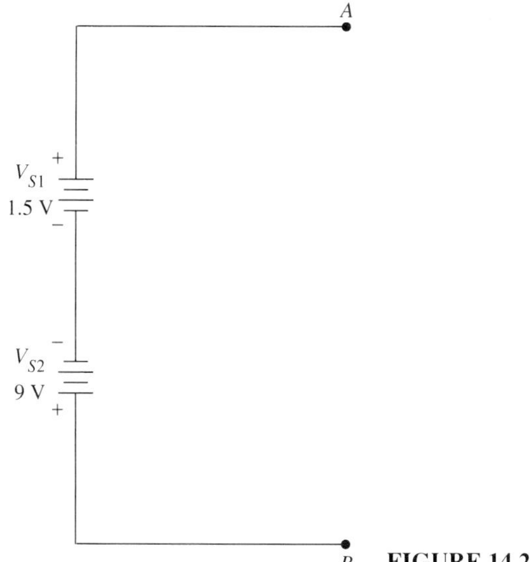

FIGURE 14.2 Series-opposing circuit.

5. Here is your chance to be creative. Design a series-aiding 15 V source with the following sources: one 9 V battery, two 1.5 V batteries, and one 12 V variable power supply. Draw the circuit schematic below. Make sure that you mark each end of each source with the correct polarity.

Series-Opposing dc Sources

6. Refer to Figure 14.2. Connect a 9 V battery and a 1.5 V battery as shown. Calculate the voltage across nodes *A* and *B*. Record your calculations in Table 14.2.
7. With the DMM set for dc volts, measure the voltage across nodes *A* and *B*. Record this measured voltage in Table 14.2.
8. Compare the measured voltage to the calculated voltage and record any percent of error in Table 14.2.
9. Comment on the amount of error. ______________________________

____

TABLE 14.2 Series-Opposing Measurements

Calculated Voltage	*Measured Voltage*	*Percent of Error*

10. Design a series-opposing 9 V source with the following sources: one 12 V source, two 1.5 V sources, and one 9 V source. You must have at least two different sources. Draw your circuit schematic below. Make sure that you mark the correct polarity at each end of each source.

QUESTIONS

Series-Aiding

1. List three everyday items that use series-aiding dc sources to obtain a certain total voltage.

   a. ______________________________

   b. ______________________________

   c. ______________________________

2. Was the voltage that you measured in Step 2 what you expected? Explain.

   ____

   ____

   ____

Series-Opposing

3. Was the voltage that you measured in Step 7 what you expected? Explain.

____

____

4. If a flashlight has three batteries (1.5 V cells) and one is placed the wrong way in the flashlight, what is the total voltage available to power the flashlight bulb? ____________ Draw a schematic showing this problem. Mark the + and − side of each cell.

Exercise 15

Parallel Circuit Characteristics

OBJECTIVES

After completing this exercise, you should be able to:

1. Identify the current paths in a parallel circuit.
2. Calculate and measure the total resistance of a parallel circuit.
3. Determine the component voltages throughout a parallel circuit.
4. Determine the power relationship for a parallel circuit.

LAB PREPARATION

Review Section 6.1 of *Introductory Electric Circuits*.

DISCUSSION

In the textbook reading, you learned that current divides into individual branch currents through parallel elements. The magnitude of these currents are inversely proportional to the magnitude of the branch resistance (Ohm's law). The text also states that when two or more components are connected in parallel, the voltage across each component is equal to the voltage across all the other components. In this exercise, we will verify the current, voltage, resistance, and power relationships for a simple parallel circuit. Figure 15.1

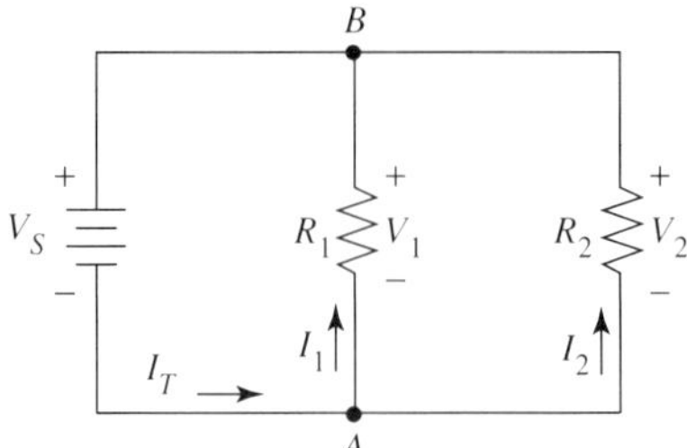

FIGURE 15.1 Simple parallel circuit.

shows a typical parallel circuit and its nomenclature. Some of the equations that will be of help are:

$$V_S = V_1 = V_2 = \ldots = V_N$$

$$I_N = \frac{V_S}{R_N}$$

$$I_S = I_1 + I_2 + \ldots + I_N$$

$$G_T = G_1 + G_2 + \ldots + G_N$$

$$R_T = \frac{1}{G_T}$$

$$R_T = \frac{V_S}{I_S}$$

$$P_T = P_1 + P_2 + \ldots + P_N$$

MATERIALS

1 dc power supply
1 DMM
1 2.2 kΩ, 0.5 W resistor
1 3.3 kΩ, 0.5 W resistor
1 set of test leads
1 protoboard

PROCEDURE

1. Measure the resistors and record their values in Table 15.1.
2. Construct the circuit in Figure 15.2. Do not connect the source voltage.
3. Predict the total resistance, R_T, of the circuit using conductance, G_T. Record both values in Table 15.2.

TABLE 15.1 Measured Resistance

Resistors	*Measured Values*
2.2 kΩ	
3.3 kΩ	

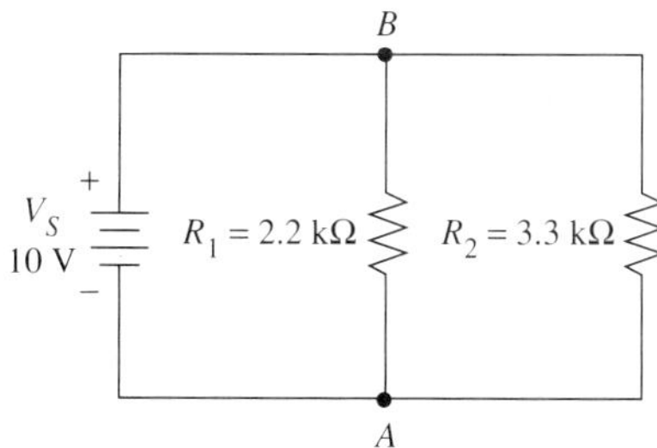

FIGURE 15.2 Parallel test circuit.

4. With the source voltage disconnected, measure the resistance of the circuit from node *A* to node *B*. Record this value in Table 15.2.

TABLE 15.2 Total Circuit Resistance

Circuit Resistance
$G_{T(CAL)} =$
$R_{T(CAL)} =$
$R_{T(MEAS)} =$

5. With the source voltage V_S connected and set to 10 V, measure the voltage drops across R_1 and R_2. Record the measured voltages in Table 15.3.

TABLE 15.3 Circuit Voltages

Measured Voltage
$V_S =$
$V_1 =$
$V_2 =$

6. Predict the source current, I_S, of the circuit. Predict the individual resistor currents of the circuit and record these three values in Table 15.4.
7. Measure the source current, I_S, by replacing the lead between the positive terminal of the voltage source and node *A* with your current meter. Record this reading in Table 15.4.

TABLE 15.4 Parallel Circuit Currents

Calculated Current	*Measured Current*
$I_S =$	$I_S =$
$I_1 =$	$I_1 =$
$I_2 =$	$I_2 =$

8. Measure the current through each branch, I_1 and I_2, of the parallel circuit. Make sure that you have completely isolated each resistor from the other when you take these current readings. Refer to Figure 15.3 for this step. Record these two readings in Table 15.4.

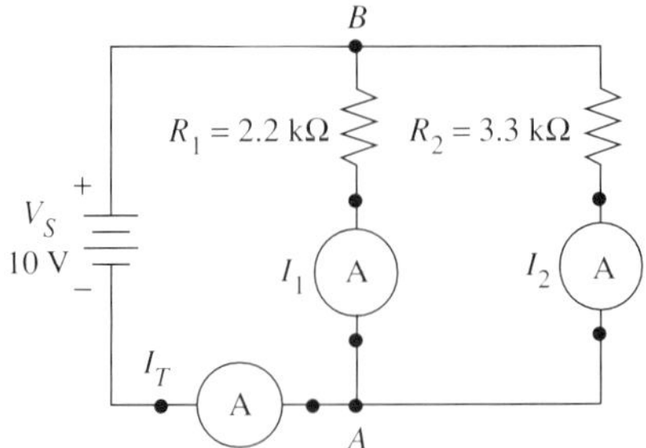

FIGURE 15.3 Parallel branch currents.

9. Predict the total circuit power, P_S. Record this value in Table 15.5.
10. Predict the power, P_1 and P_2, dissipated by each of the resistors, R_1 and R_2. Record these two values in Table 15.5.

TABLE 15.5 Calculated Parallel Power

Calculated Power
$P_S =$
$P_1 =$
$P_2 =$

QUESTIONS

1. From Table 15.2, is the calculated total resistance approximately equal to the measured resistance? ____________________

2. Is the total resistance of the circuit less than half the ohmic value of the smaller resistor value? ____________________

3. In Table 15.3, the measured values are the same magnitude for V_S, V_1, and V_2. In your own words, explain why these readings are the same. ____________________

____

____

____

4. Were the calculated currents of Table 15.4 approximately equal to the measured values? ____________________

5. Was the sum of the measured values of I_1 and I_2 approximately equal to the measured value of the source current? ____________________

6. Explain what happens to the current in a parallel circuit after it leaves the source. ____________________

____

____

7. From Table 15.5, was the power delivered to the circuit from the source dissipated by the individual resistors of the circuit? ________________

8. Write an equation to describe the relationship of Question 7. ________________

____

9. In your own words, state the characteristics that describe a parallel system. __

____

____

____

____

____

Exercise 16

Parallel Circuits: Kirchhoff's Current Law

OBJECTIVE

After completing this exercise, you should be able to:

1. Apply Kirchhoff's current law to a parallel circuit configuration.

LAB PREPARATION

Review Section 6.2 of *Introductory Electric Circuits.*

DISCUSSION

In earlier exercises, you were shown that current is constant throughout a series circuit. You were also shown that the source current in a parallel circuit equals the sum of the branch currents. In this exercise, we will examine Kirchhoff's current law. This law states that the algebraic sum of the currents entering and leaving a node must equal zero. Typically, this relationship is expressed using the following equations:

$$I_S = I_1 + I_2 + \ldots + I_N \qquad I_S - I_1 - I_2 - \ldots - I_N = 0$$

MATERIALS

1 dc power supply
1 DMM

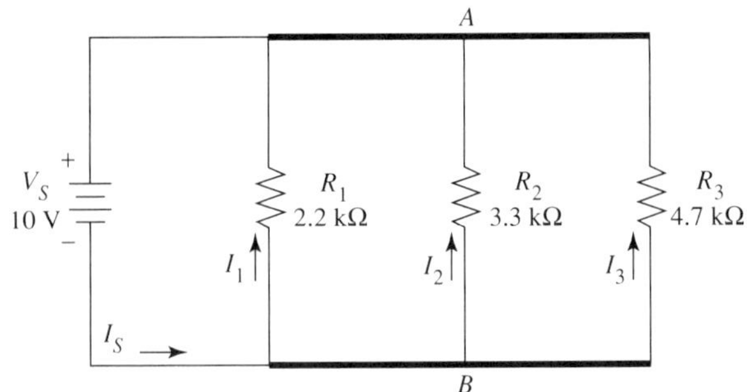

FIGURE 16.1 Kirchhoff's current law test circuit.

1 2.2 kΩ, 0.5 W resistor
1 3.3 kΩ, 0.5 W resistor
1 4.7 kΩ, 0.5 W resistor
1 set of test leads
1 protoboard

PROCEDURE

1. Measure all the resistors and record their values in Table 16.1.

TABLE 16.1 Measured Resistors

Resistors	*Measured Resistors*
$R_1 = 2.2\ \text{k}\Omega$	
$R_2 = 3.3\ \text{k}\Omega$	
$R_3 = 4.7\ \text{k}\Omega$	

2. Calculate the value of R_T. Also calculate the value of I_S, I_1, I_2, and I_3. Record these values in Table 16.2.

TABLE 16.2 Calculated and Measured Circuit Values

Resistors	*Calculated Values*	*Measured Values*
	$R_T =$	$R_T =$
	$I_T =$	$I_T =$
$R_1 = 2.2\ \text{k}\Omega$	$I_1 =$	$I_1 =$
$R_2 = 3.3\ \text{k}\Omega$	$I_2 =$	$I_2 =$
$R_3 = 4.7\ \text{k}\Omega$	$I_3 =$	$I_3 =$
Totals	$I_1 + I_2 + I_3 =$	$I_1 + I_2 + I_3 =$

3. Construct the circuit in Figure 16.1. Set the source voltage at 10 V.
4. Turn off the source with adjustments just as they are. Insert the current meter between node B and the negative terminal of the voltage source. Make sure that the meter polarity is correct. Turn the power supply back on. Measure the source current, I_S. Record this value in Table 16.2.
5. Turn the source voltage off and restore the circuit to its original condition.
6. Insert the current meter between node B and the bottom lead of R_1 (2.2 kΩ). Now, restore power to the circuit and record the value of I_1 in Table 16.2.
7. Turn off the voltage source and restore the circuit to its original condition.
8. Insert the current meter between node B and the bottom lead of R_2. Turn on the source voltage and read the current, I_2. Record this value in Table 16.2.
9. Prepare the test circuit to read the current value of I_3. Record this value in Table 16.2.

QUESTIONS

1. When you compare the calculated values of the branch currents in Table 16.2 with the measured values of the branch currents in Table 16.2, are they approximately equal? ____________________

2. Is the sum of the *calculated* currents equal to the measured value of the total source current, I_S, of Table 16.2? ____________________

3. Is the sum of all the *measured* branch currents equal to the measured value of the total source current, I_S? ____________________

4. In your own words, explain how the data in Table 16.2 supports Kirchhoff's current law. ____________________

____

____

____

5. Using the data values from Table 16.2, write two numerical forms of Kirchhoff's current law.

   a. ____________________

   b. ____________________

Exercise 17

Parallel Circuits: Current Dividers

OBJECTIVES

After completing this exercise, you should be able to:

1. Construct a current source.
2. Calculate branch currents using the current divider equation.

LAB PREPARATION

Review Section 6.3 of *Introductory Electric Circuits.*

DISCUSSION

In this exercise, you will be introduced to the current divider rule and its formulas. The current divider rule states simply that the total current entering any node will divide into as many branch currents as there are branches connected to that node. It also states that the sum of all the branch currents entering a node will be equal to the sum of the branch currents leaving that node. Below are the equations that you will need for this exercise.

$$I_1 = \frac{R_2 \times I_S}{R_1 + R_2} \qquad I_2 = \frac{R_1 \times I_S}{R + R_2} \qquad I_X = I_S \frac{R_T}{R_X}$$

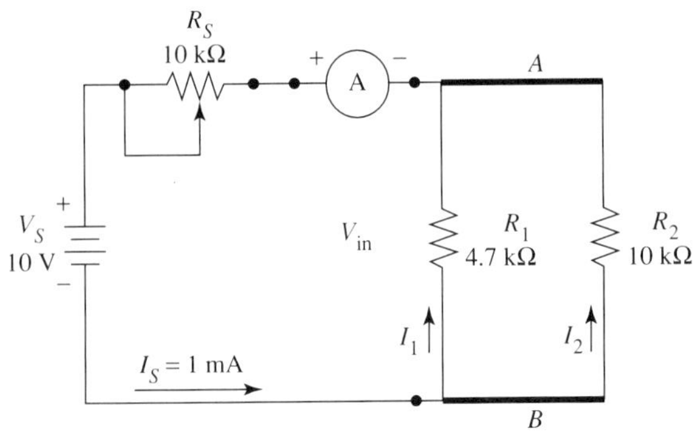

FIGURE 17.1 Current divider test circuit.

MATERIALS

1 dc power supply
1 DMM
1 VOM
1 3.3 kΩ, 0.5 W resistor
1 10 kΩ, 0.5 W resistor
1 4.7 kΩ, 0.5 W resistor
1 10 kΩ potentiometer

PROCEDURE

TABLE 17.1 Measured Resistors

Resistors	*Measured Resistors*
3.3 kΩ	
4.7 kΩ	
10 kΩ	

1. Measure and record the resistor values in Table 17.1.
2. Construct the circuit in Figure 17.1. (The 10 kΩ potentiometer will be used as a rheostat in series with the voltage source to set the current level at 1 mA.)
3. Using the current divider equations, calculate the currents through R_1 and R_2. Record these values in Table 17.2.

TABLE 17.2 Current Divider Values

Resistors	*Calculated Currents*	*Measured Currents*	*Percent of Error*
4.7 kΩ	$I_1 =$	$I_1 =$	
10 kΩ	$I_2 =$	$I_2 =$	
Total I_S	$I_S =$	$I_S =$	

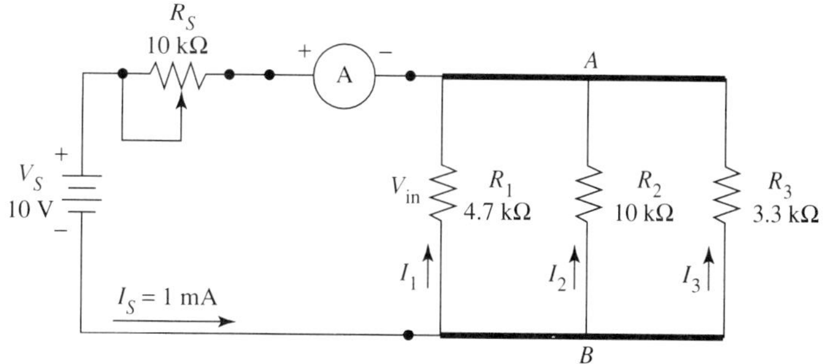

FIGURE 17.2 Current divider test circuit.

TABLE 17.3 Current Divider Values

Resistors	*Calculated Currents*	*Measured Currents*	*Percent of Error*
4.7 kΩ	I_1 −	I_1 =	
10 kΩ	I_2 =	I_2 =	
3.3 kΩ	I_3 =	I_3 =	
Total I_S	I_S =	I_S =	

4. With your DMM, set the voltage source at 10 V. Now insert the VOM as a current meter, as indicated in Figure 17.1. Adjust the rheostat until you have set the current level at 1 mA.
5. Measure the current through R_1 by inserting the DMM as a current meter in series from node *B* to the bottom lead of R_1. Record this value in Table 17.2.
6. Move the ammeter to read the current through R_2. Record this current value in Table 17.2.
7. Now turn off the source voltage and insert R_3 (3.3 kΩ) in parallel with the other two resistors. Refer to the circuit in Figure 17.2.
8. Turn on the voltage source, which is set for 10 V, and adjust the rheostat for a current of $I_S = 1$ mA.
9. Use the current divider equation to calculate the individual currents through the three resistors. Record the calculated current values in Table 17.3. I_S is the combination of the three currents.
10. Move the current meter so that you can read the current through R_1. Record this value in Table 17.3.
11. Place the current meter in series with each of the other two resistors, and measure their currents, I_2 and I_3. Record these two values in Table 17.3.

QUESTIONS

1. Explain why we use a current source in this exercise. ______________________

____

____

2. Refer to Table 17.2. Were the measured values of the two branch currents approximately equal to the calculated values of I_1 and I_2? ______________________

3. What was the largest error in your measurements? ______

4. What do you think is the reason for this error in your measurements?

______

______

______

5. Refer to Table 17.3. Were the measured values approximately equal to the calculated values? ______

6. Was the sum of the three branch currents in Table 17.3 equal to the 1 mA source current? ______

7. Using the values of Table 17.3 and the circuit of Figure 17.2, write a statement defining the current divider rule. ______

______

______

______

Exercise 18

Parallel Circuits: Fault Symptoms

OBJECTIVES

After completing this exercise, you should be able to:

1. Determine the symptoms of a parallel circuit due to different types of faults.
2. Determine the effect of an open resistor in a parallel circuit.
3. Determine the effect of a shorted resistor in a parallel circuit.

LAB PREPARATION

Review Section 6.3 of *Introductory Electric Circuits*, starting with *Fault Systems: An Open Branch.*

DISCUSSION

Figure 18.1 shows a three-branch parallel circuit with an added *series resistor* (R_S). The purpose of this resistor is to limit the current produced when a shorted branch is simulated later in the exercise. For circuit analysis purposes, R_S is considered to be part of the power supply. That is, we are using it as a variable *source resistance*. Therefore, the voltage across the parallel circuit is measured from node A to node D.

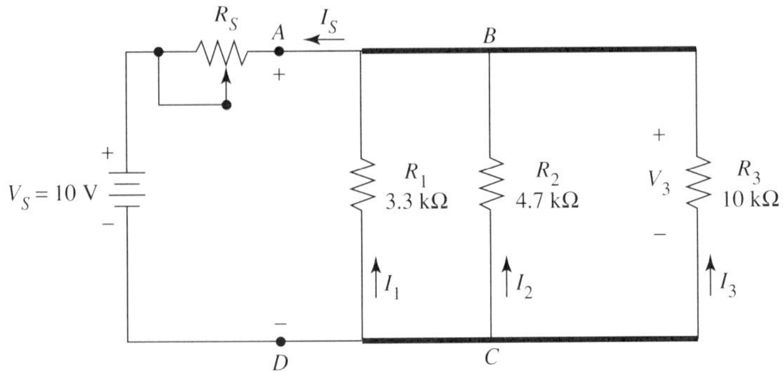

FIGURE 18.1 Fault circuit.

MATERIALS

1 dc power supply
1 DMM
1 VOM
1 3.3 kΩ, 0.5 W resistor
1 4.7 kΩ, 0.5 W resistor
1 10 kΩ, 0.5 W resistor
1 10 kΩ potentiometer
1 protoboard

PROCEDURE

1. Measure all the resistors and record their values in Table 18.1.
2. Construct the circuit in Figure 18.1. The 10 kΩ potentiometer will be used to set the circuit current and as a current limiting resistor.
3. With the DMM, set the source voltage to 10 V.
4. Insert a current meter between node *A* and node *B*, and set the source current, I_S, to 1 mA by adjusting the 10 kΩ potentiometer. Record the value of I_S in Table 18.2.
5. Now measure the source voltage at V_{AD}. Also measure the voltage drop across R_3. Record these values in Table 18.2.
6. Measure the currents in the three parallel branches and record these normal currents in Table 18.2.
7. Remove the 3.3 kΩ resistor, R_1, from the circuit by pulling the resistor from the protoboard or breadboard that you are using. This step will simulate an open circuit. Will there be any current through branch 1? ______________________

TABLE 18.1 Measured Resistors

Resistors	*Measured Values*
3.3 kΩ	
4.7 kΩ	
10 kΩ	

TABLE 18.2 Fault Value Measurements

Condition	V_{AD}	I_S	I_1	I_2	I_3	V_3
Normal						
3.3 kΩ, open						
3.3 kΩ, shorted						

8. Measure V_{AD}, V_3, I_S, I_1, I_2, and I_3 and record these values in Table 18.2.
9. Replace R_1 into the circuit and then place a jumper wire across R_1 to simulate a short circuit. **Note: If you did not have a current limiting resistor in the circuit at this time, you could damage your current meter and power source.** Will there be any voltage drop across R_1? ____________
10. Measure V_{AD}, I_S, V_3, I_1, I_2, and I_3 and record these values in Table 18.2.

QUESTIONS

1. Explain why you included R_S when you measured the source current in Step 9.

____

____

____

2. When we discover an open resistor in one of the branches of a parallel circuit, what happens to the following readings?

   a. I_S ____________

   b. V_{AD} ____________

   c. V_3 ____________

   d. I_1 ____________

   e. I_2 ____________

   f. I_3 ____________

3. Why did the value of I_S change with an open resistor in the parallel circuit? ____

____

____

____

4. Why did V_3 change value when resistor R_1 became open? ____________

____

____

5. Why did the current I_2 change value when the resistor R_1 became an open circuit? ____________

____

6. What was the value of V_{AD} when the resistor R_1 was shorted?

V_{AD} = ____________ Why? ______________________________

____

____

7. What was the voltage drop across the 10 kΩ resistor, R_3, with the shorted resistor? V_3 = ____________ Why? ______________________________

____

____

8. What was the value of I_1, as compared to I_S, with the shorted resistor?

I_1 = ____________ Why? ______________________________

____

____

9. What was the value of I_2 with the resistor shorted?

I_2 = ____________ Why? ______________________________

____

____

10. Explain what happens to the total resistance, the source current, and the voltage drop across the parallel resistance combination when a resistor becomes open in a parallel circuit configuration. ______________________________

____

____

____

11. Explain what happens to the total resistance, the source current, and the voltage drop across the parallel resistance combination when a resistor becomes shorted in a parallel circuit configuration. ______________________________

____

____

____

Exercise 19

Series-Parallel Circuits

OBJECTIVES

After completing this exercise, you should be able to:

1. Predict and measure any voltage in a series-parallel circuit.
2. Predict and measure any current through any resistor, or series grouping of resistors, in a series-parallel circuit.
3. Identify series and parallel components in a series-parallel circuit.

LAB PREPARATION

Review Sections 7.1 and 7.2 of *Introductory Electric Circuits*.

DISCUSSION

You have been taught all the current and voltage relationships for both series and parallel circuits. These relationships hold true in a series-parallel circuit. For example:

1. If two or more components are in series, the series relationships will hold true.
2. If two or more components are in parallel, the parallel relationships will hold true.

MATERIALS

1 dc power supply
1 DMM
1 200 Ω, 0.5 W resistor
1 1 kΩ, 0.5 W resistor
1 1.8 kΩ, 0.5 W resistor
1 3.3 kΩ, 0.5 W resistor
1 4.7 kΩ, 0.5 W resistor
1 protoboard
2 sets of test leads

PROCEDURE

TABLE 19.1 Measured Resistor Values

Nominal Values	*Measured Values*
$R_1 =$	
$R_2 =$	
$R_3 =$	
$R_4 =$	
$R_5 =$	

1. Measure your resistors and record their values in Table 19.1.
2. Construct the test circuit in Figure 19.1.
3. Calculate the voltages and currents listed in Table 19.2 using the nominal component values.
4. Apply power to the circuit. Measure the following values. Also, record the points in the circuit where you measured each current.

$V_1 =$ ______ $V_2 =$ ______ $V_3 =$ ______ $V_4 =$ ______ $V_5 =$ ______

$I_1 =$ ______________ Where measured: ______________

$I_2 =$ ______________ Where measured: ______________

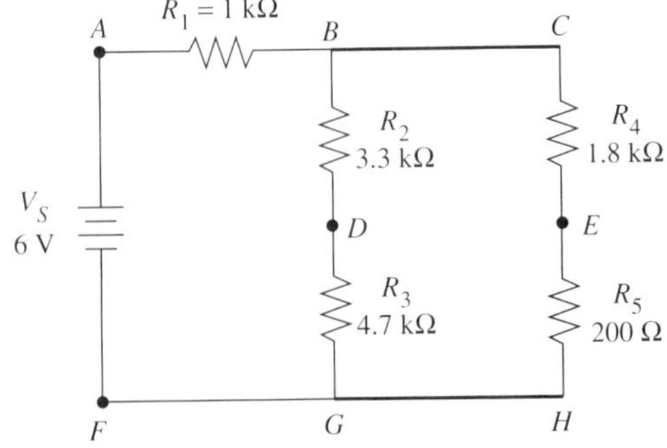

FIGURE 19.1 Series-parallel test circuit.

TABLE 19.2 Calculated Circuit Values

Resistor	*Calculated Voltage*	*Calculated Current*
$R_1 = 1\ k\Omega$	$V_1 =$	$I_1 =$
$R_2 = 3.3\ k\Omega$	$V_2 =$	$I_2 =$
$R_3 = 4.7\ k\Omega$	$V_3 =$	$I_3 =$
$R_4 = 1.8\ k\Omega$	$V_4 =$	$I_4 =$
$R_5 = 200\ \Omega$	$V_5 =$	$I_5 =$

$I_3 =$ ______________ Where measured: ______________

$I_4 =$ ______________ Where measured: ______________

$I_5 =$ ______________ Where measured: ______________

5. Using your measured current values, determine the value of I_S for the circuit.

   $I_S =$ ______________ What current values did you use to determine I_S? ______________

6. Using your values of V_S and I_S, calculate the value of R_T for the circuit.

   $R_{T(CAL)} =$ ______________

7. Disconnect the power source from your circuit and measure the value of R_T. $R_{T(MEAS.)} =$ ______________ Across what nodes did you measure to get this value of R_T? ______________

QUESTIONS/PROBLEMS

1. Using at least two sets of voltage measurements, write an equation to prove the following statement true: Series voltages add up to equal the source voltage. ______________

2. Using at least two sets of current measurements, write an equation to prove the following statement true: Current is the same through series components. ______________

3. Using circuit measurements, write an equation to prove the following statement true: Parallel voltages are equal. ______________

4. Using circuit measurements, write an equation to prove the following statement true: Parallel currents add up to equal the total current entering and leaving parallel branches at common nodes. ______________

5. Using your circuit measurements, determine whether each of the following statements is true or false. Support your discussion with a statement. As an example, the first statement is completed for you.
   a. R_2 and R_4 are in parallel. False. The voltage across the two components (V_2 and V_4) are not equal.
   b. R_2 is in series with R_3. ______
   c. (R_2 and R_3) is in parallel with (R_4 and R_5). ______
   d. R_1 is in series with $(R_2 + R_3) \| (R_4 + R_5)$. ______
   e. R_4 is in series with R_1. ______
   f. R_5 is in series with R_4. ______
   g. $I_T = I_2 + I_5$. ______
   h. R_1 is in series with the source voltage. ______

6. In your own words, discuss what you have learned from this exercise. ______

Exercise 20

Load Variations: dc Circuit Response

OBJECTIVES

After completing this exercise, you should be able to:

1. Predict the effect that a change in load resistance will have on load voltage.
2. Predict the effect that a change in load resistance will have on load current.

LAB PREPARATION

Review Section 7.2 of *Introductory Electric Circuits.*

DISCUSSION

When we speak of the *circuit response* of a dc circuit to load variations, we are talking about the changes in load voltage and current that take place when the load varies. In this exercise, we will observe the circuit response of the series-parallel circuit shown in Figure 20.1.

MATERIALS

1 dc power supply
1 DMM
1 560 Ω, 0.5 W resistor
1 1 kΩ, 0.5 W resistor

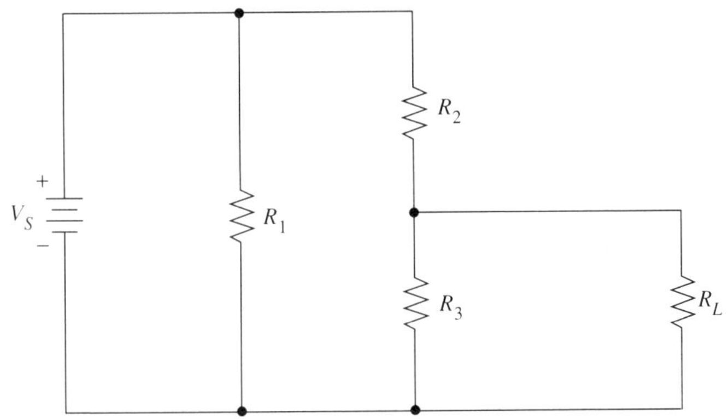

FIGURE 20.1 Simple series-parallel variable load circuit.

2 1.5 kΩ, 0.5 W resistor
1 3 kΩ, 0.5 W resistor
1 10 kΩ, 0.5 W resistor
1 protoboard
2 sets of test leads

PROCEDURE

1. Measure all resistor values and record them in Table 20.1.
2. With the power off, construct the circuit of Figure 20.2.
3. Predict the voltage drops across R_2, R_3, and R_L. Record your predicted values in Table 20.2.
4. Turn on the voltage source. Using the DMM, set the source voltage to 8 V.
5. Measure the voltage drops across R_2, R_3, and R_L. Record these values in Table 20.3.
6. Predict the values of current through R_2, R_3, and R_L. Record these values in Table 20.2.
7. Turn off the power supply and insert the DMM as a current meter to measure I_2. Turn on the power supply and measure I_2. Record this value in Table 20.3.
8. Make sure that the voltage source is still set at 8 V. Use the same procedure as in Step 7 to measure the currents I_3 and I_L. Record these values in Table 20.3.
9. Turn off the power supply and replace the 1.5 kΩ resistor with the 500 Ω resistor.

TABLE 20.1 Measured Resistance

Resistors	*Measured Values*
$R_1 = 1$ kΩ	
$R_2 = 1.5$ kΩ	
$R_3 = 3$ kΩ	
$R_{L1} = 1.5$ kΩ	
$R_{L2} = 500$ Ω	
$R_{L3} = 10$ kΩ	

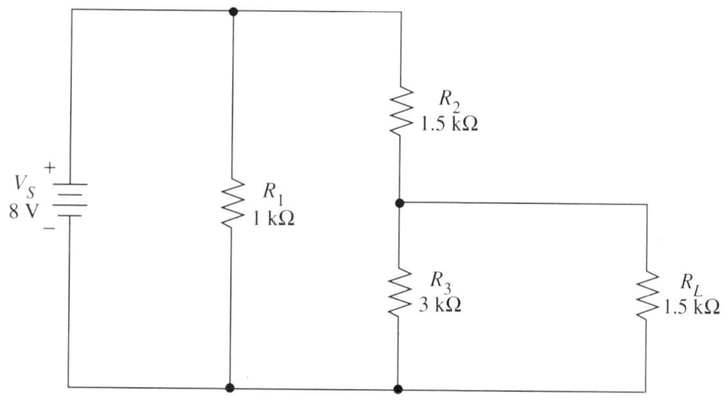

FIGURE 20.2 Variable load test circuit.

TABLE 20.2 Predicted Values

	$R_L = 1.5\ k\Omega$	$R_L = 500\ \Omega$	$R_L = 10\ k\Omega$
V_{R2}			
V_{R3}			
V_{RL}			
I_2			
I_3			
I_L			

10. Predict the values of V_2, V_3, V_L, I_2, I_3, and I_L. Record all these values in Table 20.2.
11. Measure the voltage drops across R_2, R_3, and R_L. Record these values in Table 20.3.
12. Measure the currents I_2, I_3, and I_L. Record these values in Table 20.3.

TABLE 20.3 Measured Values

	$R_L = 1.5\ k\Omega$	$R_L = 500\ \Omega$	$R_L = 10\ k\Omega$
V_{R2}			
V_{R3}			
V_{RL}			
I_2			
I_3			
I_L			

13. Turn off the power supply and construct the test circuit with $R_L = 10\ k\Omega$.
14. Repeat Steps 10 through 12 for the new load resistance. Record the appropriate predicted values in Table 20.2 and the measured values in Table 20.3.

QUESTIONS

1. Refer to Table 20.3. Did the measured voltage drops across the load resistance vary from one load resistance to the next? ____________________
   What was the percent of change for each load resistance? ____________________

   ____

2. Refer to Table 20.3. Did the currents through the load resistance vary from one load resistance to the next? ____________________
   What was the percent of change for each load resistance? ____________________

   ____

3. Refer to Table 20.3. Explain why the load voltage varied from one load resistance to the next. ____________________

   ____

   ____

4. Compare the values of Table 20.2 to the measured values of Table 20.3. Did the results come out as you expected? Why? ____________________

   ____

   ____

5. What observations can you make from this exercise?

   ____

   ____

   ____

   ____

Exercise 21

Loaded Voltage Dividers

OBJECTIVES

After completing this exercise, you should be able to:

1. Predict the effect that a change in load resistance will have on the output from a voltage divider circuit.
2. Predict the minimum value of R_L that maintains load voltage stability to within $\pm 10\%$.

LAB PREPARATION

Review Section 7.3 of *Introductory Electric Circuits.*

DISCUSSION

A voltage divider is usually designed to provide stable output voltage (or range of voltages) for a specified load. Figure 21.1 shows a simple loaded voltage divider. The load voltage for this circuit varies directly with the resistance of the load. The variation in load voltage depends on the *stability* of the voltage divider. As long as the bleeder current (I_2) is at least 10 times the load current, the voltage across the load remains relatively stable against changes in load resistance. In this exercise, we will take a closer look at voltage divider stability.

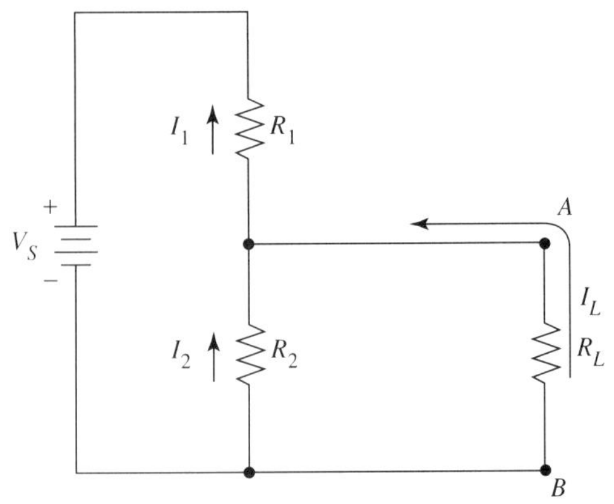

FIGURE 21.1 Simple voltage divider circuit.

MATERIALS

1 dc power supply
1 DMM
1 2.2 kΩ, 0.50 W resistor
1 5.6 kΩ, 0.5 W resistor
1 15 kΩ, 0.5 W resistor
1 33 kΩ, 0.5 W resistor
1 68 kΩ, 0.5 W resistor
1 100 kΩ, 0.5 W resistor
1 protoboard
2 sets of test leads

PROCEDURE

TABLE 21.1 Measured Resistors

Resistors	*Measured Values*
$R_1 = 5$ kΩ	
$R_2 = 2.2$ kΩ	
$R_{L1} = 100$ kΩ	
$R_{L2} = 67$ kΩ	
$R_{L3} = 33$ kΩ	
$R_{L4} = 15$ kΩ	

1. Measure the resistors and record their values in Table 21.1.
2. Construct the voltage divider in Figure 21.2.
3. Predict the values of no-load output voltage across nodes A and B and bleeder current (I_2).

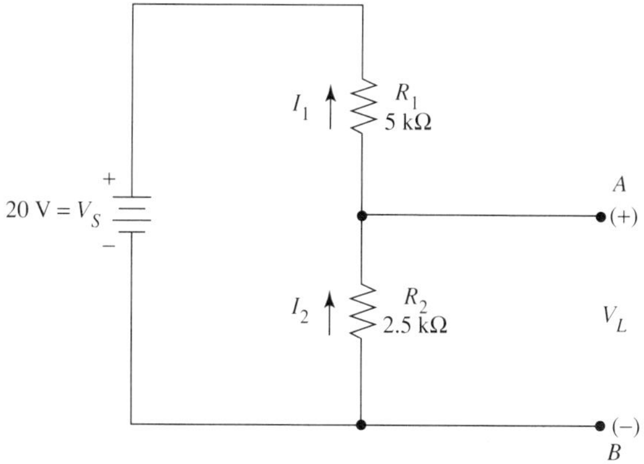

FIGURE 21.2 Voltage divider test circuit (no-load).

4. Using the DMM, set the voltage source for a 20 V output.
5. Measure the no-load output voltage and the bleeder current. Record these readings in Table 21.2.
6. Add the load shown in Figure 21.3 to the circuit.
7. Predict the values of load voltage and current. Record these values in Table 21.3.
8. Measure V_L and I_L. Record these values in Table 21.3.
9. Replace the 100 kΩ resistor with the following resistors, one at a time: 67 kΩ, 33 kΩ, 15 kΩ. Repeat Steps 7 and 8 for each resistor. Record all your predicted and measured values in Table 21.3.

TABLE 21.2 No-Load Circuit Measurements

No-Load Calculations	*No-Load Measurements*
$V_2 = V_L =$	$V_2 = V_L =$
$I_1 = I_2 =$	$I_1 = I_2 =$

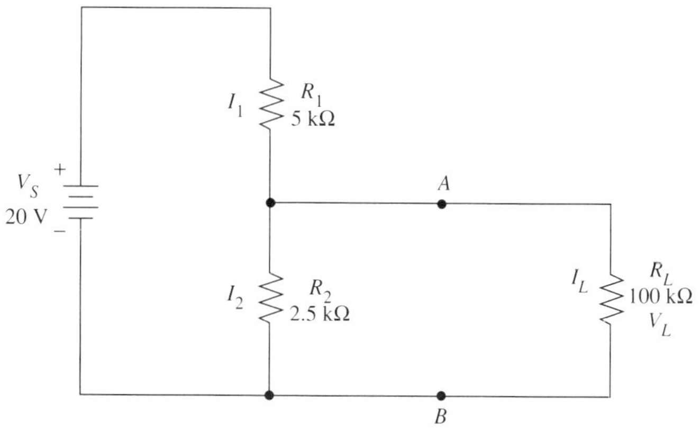

FIGURE 21.3 Voltage divider test circuit (loaded).

TABLE 21.3 Loaded Voltage Divider Measurements

Load Resistors	V_L *Calculated*	I_L *Calculated*	V_L *Measured*	I_L *Measured*
100 kΩ				
67 kΩ				
33 kΩ				
15 kΩ				

QUESTIONS

1. Refer to Table 21.2. Did your predicted values equal your measured values? If not, why? ____________
2. Refer to Table 21.3. Did your predicted values equal your measured values? If not, which ones were off and why? ____________
3. Refer to the values in Table 21.3. What was the difference between the values of V_L for each load resistor and the no-load value of V_L? Give each voltage and its percent of error for each value of R_L:

   No-load V_L = ________ Percent of error ________

   100 kΩ V_L = ________ Percent of error ________

   67 kΩ V_L = ________ Percent of error ________

   33 kΩ V_L = ________ Percent of error ________

   15 kΩ V_L = ________ Percent of error ________
4. Refer to the information in Question 3. What would be the smallest resistor value that we could use to maintain at least a 10% drop in load voltage?

   ____
5. In your own words, explain what happens to the load voltage as the load resistance decreases. ____________

   ____

   ____

Exercise 22

The Wheatstone Bridge

OBJECTIVES

After completing this exercise, you should be able to:

1. Describe the construction and operation of a Wheatstone bridge.
2. Use a Wheatstone bridge as a resistance measuring device.

LAB PREPARATION

Review Section 7.4 of *Introductory Electric Circuits.*

DISCUSSION

The Wheatstone bridge is a very specific type of circuit. Figure 22.1 shows a Wheatstone bridge. It has a dc voltage source, four essential resistances, and a voltage galvanometer connected between node A and node B. Some equations that will be useful for this exercise are:

$$\frac{R_2}{R_1} = \frac{R_4}{R_3} \qquad R_X = R_{db} \times \frac{R_3}{R_4}$$

In the second equation, if the ratio of R_3 to R_4 is equal to 1, then the value of the unknown resistor (R_X) can be read from the dial positions of a resistance decade box (R_{db}).

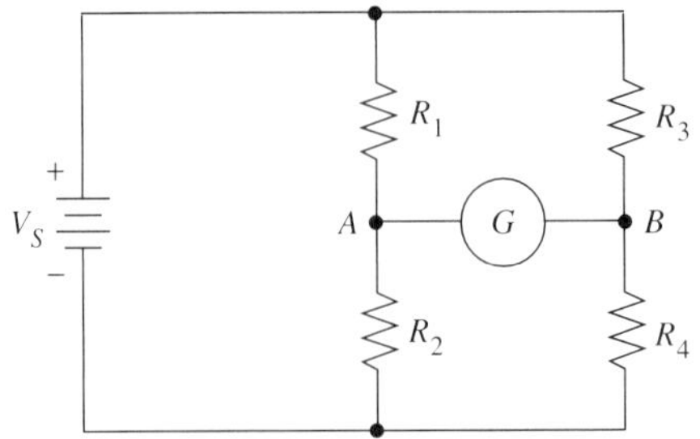

FIGURE 22.1 The Wheatstone bridge.

MATERIALS

1 dc power supply
1 DMM
1 decade resistance box
1 1 kΩ, 0.5 W resistor
2 2 kΩ, 0.5 W resistors
1 3.3 kΩ, 0.5 W resistor
1 4.7 kΩ, 0.5 W resistor
1 15 kΩ potentiometer
1 protoboard
2 sets of test leads

PROCEDURE

TABLE 22.1 Measured Resistors

Resistors	*Measured Resistors*
1 kΩ	
2 kΩ	
2 kΩ	
3.3 kΩ	
4.7 kΩ	

1. Measure the resistors and record their values in Table 22.1.
2. Construct the circuit in Figure 22.2 using the appropriate resistors. You will assign R_1 as the unknown resistor, R_X, and R_2 as the variable resistor. Make sure that the potentiometer is set for maximum resistance. Use your 3.3 kΩ resistor as R_X.
3. Set the voltage source to 8 V using the DMM.
4. Set the DMM to 22 V and connect it between node *A* and node *B* of the circuit. The meter will be used in place of a galvanometer.
5. Now adjust the potentiometer until the voltage reading approaches 0 V. As the voltage decreases, decrease the voltage range setting of the DMM until you are on the lowest range. Adjust the potentiometer until you read 0 V. When the meter reads 0 V, the bridge is balanced.

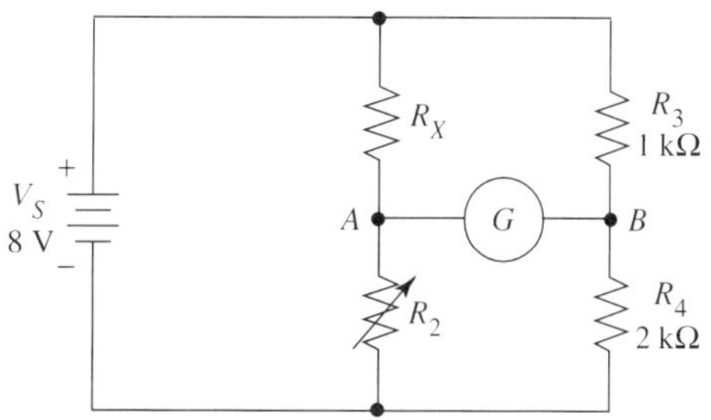

FIGURE 22.2 Wheatstone bridge test circuit.

TABLE 22.2 Measuring Unknown Resistor Values

Actual Measured Value	*Value Using Potentiometer*	*Value Using Decade Box*	*Value Using $R_3 = R_4$*
3.3 kΩ =			
4.7 kΩ =			

6. Without changing its setting, disconnect the potentiometer and measure its resistance. R_{pot} = ____________
7. Use the following equation:

$$R_X = R_1 = R_{pot} \times \frac{R_3}{R_4}$$

This equation gives you the value of the unknown resistor, R_X, when you use a potentiometer. Record the calculated value of the unknown resistor in Table 22.2.

8. Replace the 3.3 kΩ resistor with the 4.7 kΩ resistor and repeat Steps 2 through 7. Record the calculated unknown resistance in Table 22.2.
9. Replace the potentiometer with the resistance decade box and repeat Steps 2 through 8, first with the 3.3 kΩ and then with the 4.7 kΩ resistors. Use the following equation:

$$R_X = R_{db} \times \frac{R_3}{R_4}$$

Record these values in Table 22.2.

10. Replace R_3 with a 2 kΩ resistor that is as close in ohmic value as you can find to R_4. R_3 is now approximately equal to R_4.
11. Using the 3.3 kΩ resistor as R_X, adjust the decade resistance until you see a 0 V reading between node *A* and node *B*. Read the value of R_X from the dials of the decade box and record the value in Table 22.2.
12. Replace the 3.3 kΩ resistor with the 4.7 kΩ resistor and repeat Step 11. Record this value in Table 22.2.

QUESTIONS

1. How do you know when a Wheatstone bridge is balanced? ____________

____

____

2. Which of the three methods that we used gave the most accurate reading?

____

____

____

3. In Steps 11 and 12, why were you able to read the value of the unknown resistor from the dials of the decade box? ______________________

____

____

4. When using a Wheatstone bridge to measure resistance, the accuracy of the reading depends on what two factors?

   a. ____

   b. ____

5. When using a Wheatstone bridge to measure an unknown resistance, why is it important that the bridge be in balance? ______________________

____

____

Exercise 23

Superposition

OBJECTIVES

After completing this exercise, you should be able to:

1. Analyze a two-source circuit using the superposition theorem.
2. Measure the voltages and currents to demonstrate the superposition theorem.

LAB PREPARATION

Review Section 8.1 of *Introductory Electric Circuits.*

DISCUSSION

Some circuits contain two or more voltage sources and/or two or more current sources. These types of circuits are called *multisource* circuits. The superposition theorem states that the response of a circuit to more than one source (as shown in Figure 23.1) can be determined by analyzing the circuit response to each source alone and then combining the results.

MATERIALS

1 dc power supply
1 DMM

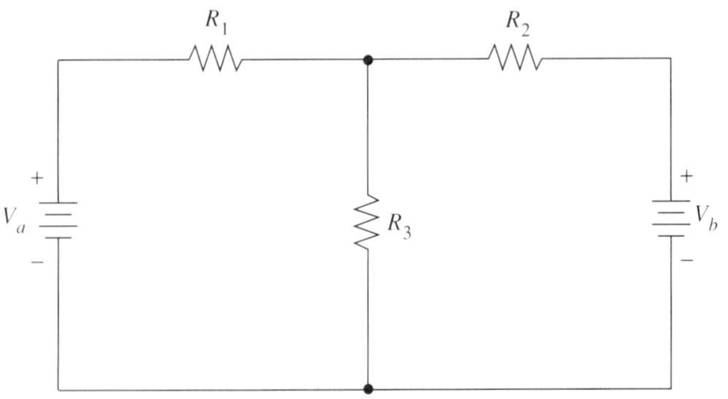

FIGURE 23.1 Multisource circuit.

1 100 Ω, 0.5 W resistor
1 150 Ω, 0.5 W resistor
1 22 Ω, 0.5 W resistor
1 protoboard
2 sets of test leads

PROCEDURE

1. Measure the resistors and record their values in Table 23.1.
2. Construct the circuit in Figure 23.2.
3. Turn off V_B by disconnecting the source from the circuit. Create a short between the two terminals where the source, V_B, had been connected. Set the voltage source V_A to 12 V as shown in Figure 23.3.

TABLE 23.1 Measured Resistor Values

Nominal Values	*Measured Values*
$R_1 = 100\ \Omega$	
$R_2 = 22\ \Omega$	
$R_3 = 150\ \Omega$	

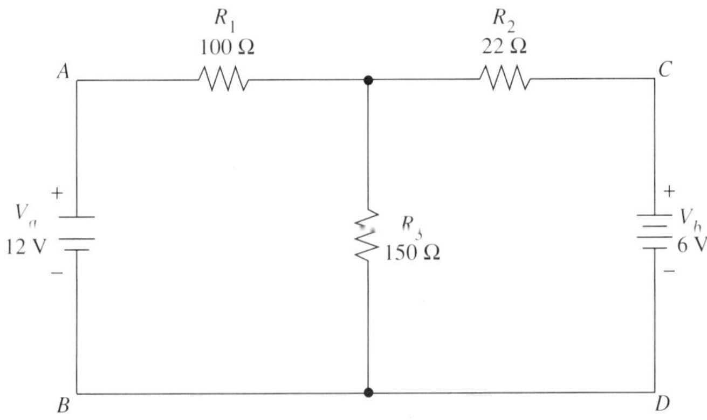

FIGURE 23.2 Superposition test circuit.

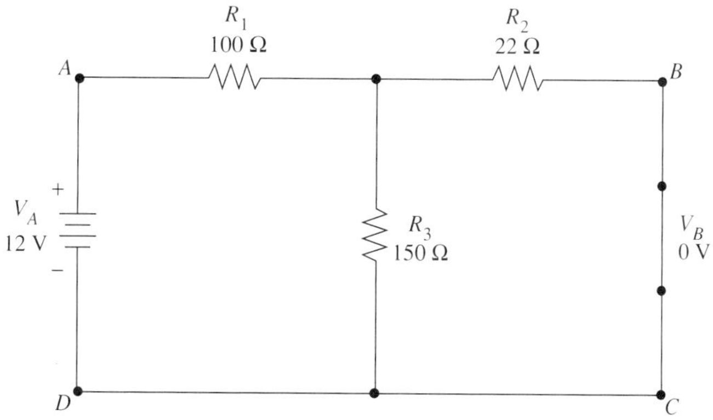

FIGURE 23.3 Circuit analysis for V_A.

4. Predict and measure the values of the voltage and current for the source V_A from Figure 23.3. As you solve the circuit, remember to keep track of the polarities of the voltage drops and the directions of the currents generated by the source. Record the voltages and the currents in Table 23.2. Don't forget to indicate the polarities and current directions. You can do this by marking Figure 23.3.
5. Reconstruct the circuit of Figure 23.2. Now turn off V_A. Disconnect the voltage source, V_A, and create a short between the open terminals where V_A had been. Set the voltage V_B to 6 V, as shown in Figure 23.4.
6. Predict and measure the voltage drops and currents for the voltage source V_B as shown in Figure 23.4. Remember to keep track of the polarities and current directions for the predicted and measured values. Record these values in Table 23.3. Mark the polarities and current directions in Figure 23.4.
7. Determine the sum of the predicted voltages and currents listed in Tables 23.2 and 23.3. Record these sums in Table 23.4.

TABLE 23.2 Predicted and Measured Values for V_A

Resistors	*Predicted Voltage*	*Predicted Current*	*Measured Voltage*	*Measured Current*
R_1				
R_2				
R_3				

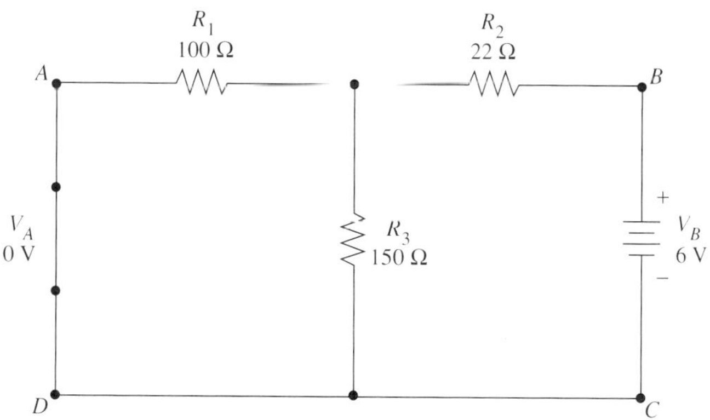

FIGURE 23.4 Circuit analysis for V_B.

TABLE 23.3 Predicted and Measured Values for V_B

Resistors	*Predicted Voltage*	*Predicted Current*	*Measured Voltage*	*Measured Current*
R_1				
R_2				
R_3				

8. Reconstruct the circuit of Figure 23.2. Now measure the voltages and currents and their polarities and direction for each of the resistors of the circuit. Record these current values and directions in Table 23.4.

TABLE 23.4 Combined Predicted and Measured Values of Voltage and Current

Resistors	*Predicted Voltage*	*Predicted Current*	*Measured Voltage*	*Measured Current*
R_1				
R_2				
R_3				

QUESTIONS

1. Refer to Table 23.4. Did the measured voltage values and polarities of the circuit coincide with the predicted values of the circuit? ____________________
2. Refer to Table 23.4. Did the measured currents and their directions coincide with the predicted values of the circuit? ____________________
3. What conclusions can you draw from this exercise?

____

____

____

Exercise 24

Thevenin's Theorem

OBJECTIVES

After completing this exercise, you should be able to:

1. Derive the Thevenin equivalent of a series or a series-parallel circuit.
2. Use the Thevenin equivalent circuit to predict the values of load voltage for different load resistor values.

LAB PREPARATION

Review Section 8.3 of *Introductory Electric Circuits*.

DISCUSSION

Thevenin's theorem states that any circuit or system can be represented as a single source voltage, V_{th}, in series with a single resistor, R_{th}, as shown in Figure 24.1. The Thevenin equivalent circuit can be used to predict how the original circuit's output values will change when the load resistor changes.

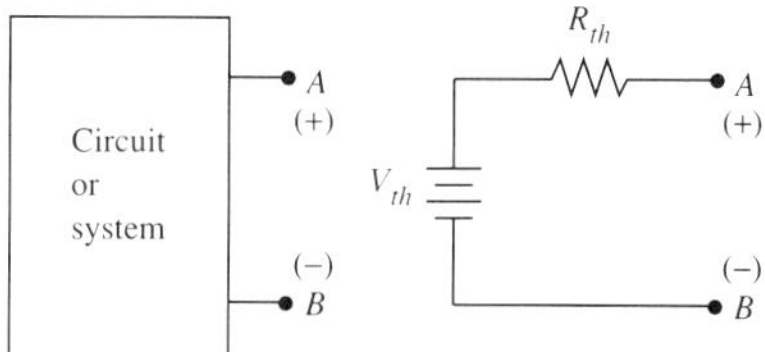

FIGURE 24.1 Thevenin equivalent circuit.

MATERIALS

1 dc power supply
1 DMM
1 decade resistance box
2 1 kΩ, 0.5 W resistor
1 2 kΩ, 0.5 W resistor
1 2.2 kΩ, 0.5 W resistor
1 3.3 kΩ, 0.5 W resistor
1 4.7 kΩ, 0.5 W resistor
1 10 kΩ, 0.5 W resistor
1 15 kΩ, 0.5 W resistor
1 protoboard

PROCEDURE

TABLE 24.1 Measured Resistors

Resistors	*Measured Values*
1 kΩ	
1 kΩ	
2 kΩ	
2.2 kΩ	
3.3 kΩ	
4.7 kΩ	
10 kΩ	
15 kΩ	

1. Derive the Thevenin equivalent values of R_{th} and V_{th} for the series circuit shown in Figure 24.2 using measured component values. Draw the Thevenin equivalent circuit neatly in the space provided above.
2. Make sure that the power is off. Construct the Thevenin equivalent circuit. Use the decade resistance box to provide the value of R_{th}.

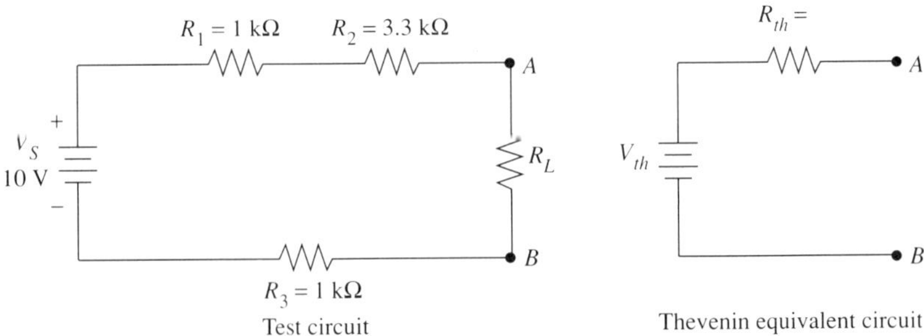

FIGURE 24.2 A series circuit and its Thevenin series equivalent.

3. For each of the measured load resistances (R_L) listed below, predict the value of the load voltage that the Thevenin equivalent circuit will provide.

   $R_L = 4.7\ \text{k}\Omega$ $V_L =$ ________

   $R_L = 10\ \text{k}\Omega$ $V_L =$ ________

   $R_L = 15\ \text{k}\Omega$ $V_L =$ ________
4. Turn on the voltage source, and set it for the value of the Thevenin voltage, V_{th}.
5. Using the DMM, measure the output voltage from the Thevenin equivalent circuit for each value of load resistance. Record the measured values in the space provided.

   $R_L = 4.7\ \text{k}\Omega$ $V_L =$ ________

   $R_L = 10\ \text{k}\Omega$ $V_L =$ ________

   $R_L = 15\ \text{k}\Omega$ $V_L =$ ________
6. Construct the original series circuit shown in Figure 24.2. For each value of load resistance, measure and record the output voltage below.

   $R_L = 4.7\ \text{k}\Omega$ $V_L =$ ________

   $R_L = 10\ \text{k}\Omega$ $V_L =$ ________

   $R_L = 15\ \text{k}\Omega$ $V_L =$ ________
7. Derive the Thevenin equivalent values of R_{th} and V_{th} for the series-parallel circuit in Figure 24.3. Draw the Thevenin equivalent circuit neatly in the space provided in the figure.
8. Construct the Thevenin equivalent circuit using the potentiometer to provide the value of R_{th}.
9. For each of the measured load resistors listed below, predict the value of the load voltage that the Thevenin equivalent circuit will provide.

   $R_L = 4.7\ \text{k}\Omega$ $V_L =$ ________

   $R_L = 10\ \text{k}\Omega$ $V_L =$ ________

   $R_L = 15\ \text{k}\Omega$ $V_L =$ ________
10. Turn on the voltage source and set it for V_{th}.
11. Using the DMM, measure the output voltage for each load resistance. Record these voltages below.

   $R_L = 4.7\ \text{k}\Omega$ $V_L =$ ________

   $R_L = 10\ \text{k}\Omega$ $V_L =$ ________

   $R_L = 15\ \text{k}\Omega$ $V_L =$ ________

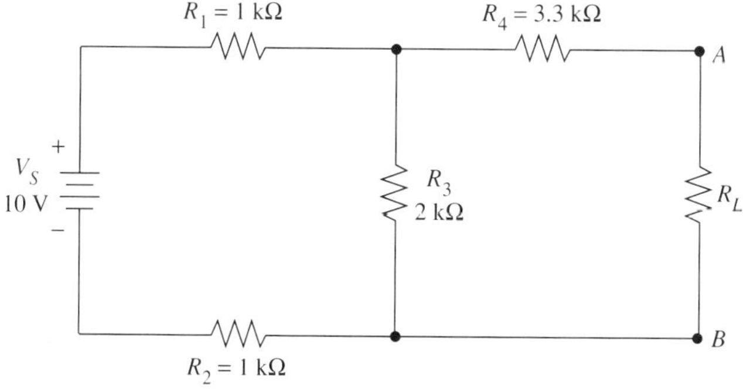

Test circuit

Thevenin equivalent circuit

FIGURE 24.3 A series-parallel circuit and its Thevenin equivalent.

12. Construct the original series-parallel circuit shown in Figure 24.3.
13. Turn on the voltage source and set V_S to 10 V.
14. For each value of load resistance, measure and record the output voltage.

$R_L = 4.7\ k\Omega$ $V_L =$ ____________

$R_L = 10\ k\Omega$ $V_L =$ ____________

$R_L = 15\ k\Omega$ $V_L =$ ____________

QUESTIONS

1. Compare the predicted voltages of Step 3 with the measured values of Step 5. What is your conclusion about the comparison of these two sets of voltages?

2. Compare the measured output voltages produced by the Thevenin equivalent circuit in Step 5 to the measured output voltages of Step 6 for the original circuit. What did you observe?

3. Compare the predicted output voltages of the series-parallel Thevenin equivalent circuit in Step 9 to the measured values of Step 11. What is your conclusion regarding these two sets of voltages?

4. Compare the measured output voltages in Step 11 to the measured output voltages in Step 14. What is your conclusion regarding these two sets of output voltages?

5. Explain what you have learned in this exercise about Thevenin equivalent circuits.

Exercise 25

Maximum Power Transfer

OBJECTIVES

After completing this exercise, you should be able to:

1. Demonstrate the relationship between maximum load power and Thevenin resistance for a circuit with a variable load.
2. Demonstrate the relationship between maximum load power and source resistance for a circuit with a variable load.

LAB PREPARATION

Review Section 8.4 of *Introductory Electric Circuits*.

DISCUSSION

In Section 5.4 you were introduced to maximum power transfer for a series circuit with a variable load. For a series-parallel circuit with a variable load, maximum power transfer occurs when the load resistance, R_L, is equal to the Thevenin resistance of the circuit. The following equations will help you complete this exercise.

$$V_L = V_{th} \times \frac{R_L}{R_L + R_{th}} \qquad P_L = \frac{V_L^2}{R_L}$$

MATERIALS

1 dc power supply
1 DMM
1 68 Ω, 0.5 W resistor
1 220 Ω, 0.5 W resistor
1 330 Ω, 0.5 W resistor
1 decade resistance box
1 protoboard
2 sets of test leads

TABLE 25.1 Measured Resistance Values

Nominal Values	*Measured Values*
$R_1 = 330\ \Omega$	
$R_2 = 220\ \Omega$	
$R_3 = 68\ \Omega$	

PROCEDURE

Maximum Power Transfer

1. Measure all resistor values and record the measurements in Table 25.1.
2. Construct the circuit in Figure 25.1. Leave out the potentiometer.
3. Predict the values of V_{th}, R_{th}, V_L, and maximum load power. Record these values in Table 25.2.
4. With the voltage source turned off and the voltage source nodes shorted together, measure the resistance across the open load terminals with the DMM. This resistance is the Thevenin resistance of the circuit (R_{th}). Record this value in Table 25.2.
5. Remove the short from across the output nodes of the power supply.
6. Turn on the power supply and, using the DMM, set the source voltage to 12 V.
7. Measure the voltage across the open terminals of the load. This voltage is the value of V_{th}. Record this value in Table 25.2.
8. Set the decade box to the measured value of R_{th} from Step 4. Insert the decade box into the circuit between the open terminals of the load. With the decade box at this setting, $R_{th} = R_L$. Record this value in Table 25.2 as R_L.

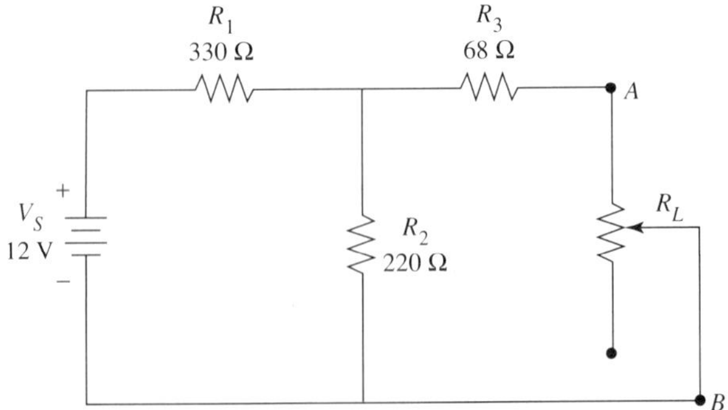

FIGURE 25.1 Maximum power transfer test circuit.

TABLE 25.2 Maximum Power Values

Values	*Predicted*	*Measured*
$R_{th} = R_L$		
V_L		
P_L		

9. Measure the voltage drop across the load resistance (V_L). Record this value in Table 25.2.
10. Using the measured values of V_L and R_L, calculate the measured value of load power. Record this value of P_L in Table 25.2.

Maximum Power Transfer Curve

1. Using the test circuit in Figure 25.1, set R_L to 20 Ω.

   Note: Do not forget to turn off the voltage source each time that you reset the resistance of R_L.

   When the resistance is set, turn on the voltage source. Make sure that your voltage source is set to 12 V. Measure the voltage drop, V_L, across R_L. Record the value of V_L in Table 25.3.
2. Repeat this sequence with the decade resistance box set for 50 Ω, 100 Ω, 150 Ω, 200 Ω, 250 Ω, 300 Ω, 350 Ω, 400 Ω, and 500 Ω. Complete Table 25.3 with the value for V_L of each resistance listed.
3. Using the values in Table 25.3, calculate the value of P_L for each set of R_L and V_L. Record these values in the table.
4. Using the data from Table 25.3, graph the maximum power transfer curve on the graph Figure 25.2. Label each point on the curve.

TABLE 25.3 Maximum Power Transfer Curve Values

Resistance	*Voltage (V_L)*	*Power (P_L)*
20 Ω		
50 Ω		
100 Ω		
150 Ω		
200 Ω		
250 Ω		
300 Ω		
350 Ω		
400 Ω		
500 Ω		

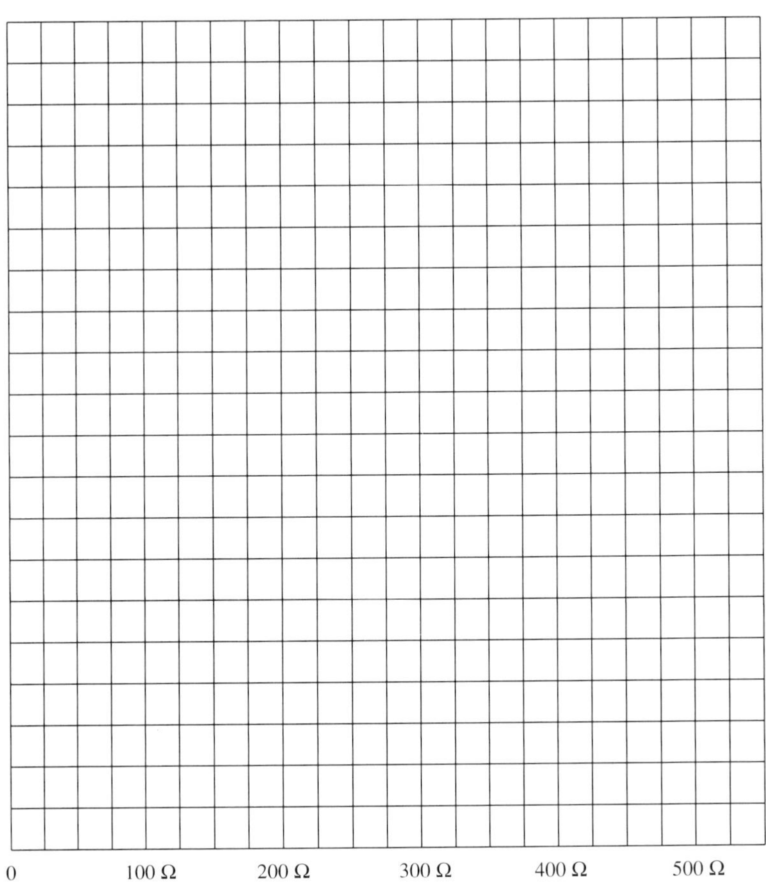

FIGURE 25.2 Maximum power transfer curve.

QUESTIONS

Maximum Power Transfer

1. Refer to Table 25.2. Did the measured values of R_{th} and V_{th} equal the calculated values? ____________________
2. What was the percent of error between the measured and the calculated values?
   a. V_{th}: Percent of error ____________________
   b. R_{th}: Percent of error ____________________
3. From what you did in Exercise 13, was the voltage source displaying maximum power transfer when R_L was equal to R_S? Why or why not? ____________________

Maximum Power Transfer Curve

1. Refer to Figure 25.2. Did the maximum point of the curve indicate (P_{max}) to be equal to the measured value of maximum load power in Table 25.2? __________

List each value:

P_{max} > Table 25.2 ____________________

P_{max} > Figure 25.2 ____________________

2. What conclusions can you draw from the results of the graphed line in Figure 25.2? ____

____

Exercise 26

Mesh Current Analysis

OBJECTIVE

After completing this exercise, you should be able to:

1. Test and solve a 2-loop multisource circuit using mesh current analysis.

LAB PREPARATION

Review Section 9.3 of *Introductory Electric Circuits*.

DISCUSSION

The branch current analysis method has its limitations and superposition is time consuming, so we need a faster way to solve multisource circuits. Mesh analysis, a more universal system of analysis, is a system for solving multisource circuits. The primary difference between branch current analysis and mesh circuit analysis is that with mesh current analysis, you will assume that each loop current is in a clockwise direction, regardless of the polarity of the voltage source in that loop. If you look at Figure 26.1, you will see that all the loop currents are drawn clockwise. As you read Section 9.3 of the textbook, you will learn how to use the mesh analysis system to solve multisource circuits.

MATERIALS

1 330 Ω, 0.5 W resistor
1 120 Ω, 0.5 W resistor

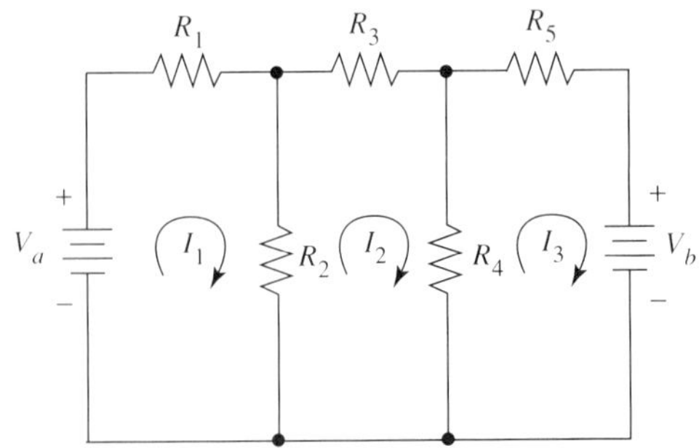

FIGURE 26.1 Mesh current in a 3-loop circuit.

1 220 Ω, 0.5 W resistor
1 dual dc power supply *or* 2 dc power supplies
1 DMM
1 protoboard

PROCEDURE

1. Measure all the resistors and record their values in Table 26.1.
2. Predict the values for I_1, I_2, and I_3 for Figure 26.1 using the mesh analysis method. Predict the values of V_1, V_2, and V_3. Do not forget what their polarities are. Once you have determined the six values, record them below.

   I_1 = ______ I_2 = ______ I_3 = ______

   V_1 = ______ V_2 = ______ V_3 = ______

   > Note: Use Figure 26.2 to mark the direction and polarities of the different voltages and currents.

3. Construct the test circuit in Figure 26.2 using the appropriate power supplies.
4. Using the DMM, set the voltage of V_A to 12 V and V_B to 8 V. Recheck V_A and V_B after they are first set.
5. Measure V_1, V_2, and V_3. Be sure to determine their polarities. Record these values and their polarities below.

   V_1 = ______ V_2 = ______ V_3 = ______
6. Measure the three currents, I_1, I_2, and I_3, and determine their directions.

   I_1 = ______ I_2 = ______ I_3 = ______

TABLE 26.1 Measured Resistor Values

Nominal Values	*Measured Values*
R_1 = 330 Ω	
R_2 = 120 Ω	
R_3 = 220 Ω	

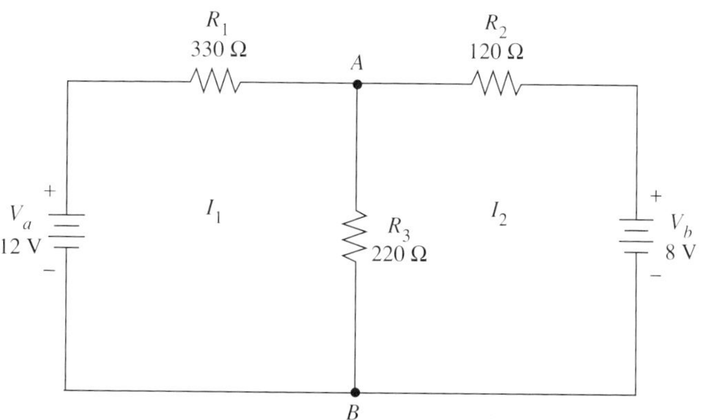

FIGURE 26.2 Mesh analysis test circuit.

QUESTIONS

1. Were the measured values of V_1, V_2, V_3, and their polarities the same as the predicted values? ____
2. Were the measured values of I_1, I_2, and their current directions the same as the predicted values? ____
3. For the voltage values that you measured and their polarities, write a Kirchhoff's voltage law equation around each loop of the circuit.

   Loop 1: ____

   Loop 2: ____
4. For the currents leaving and entering node *A*, write Kirchhoff's current law for node *A*: ____ .

Part 2

ac Exercises

Exercise 27

The Oscilloscope and Its Use

OBJECTIVES

After completing this exercise, you should be able to:

1. Measure the magnitude of a dc voltage with an oscilloscope.
2. Measure the magnitude of an ac voltage and its frequency with an oscilloscope.
3. Determine the frequency of a given sinusoidal ac waveform.
4. Measure the frequency of an unknown sinusoidal waveform.

LAB PREPARATION

Review Sections 11.1 and 11.2 of *Introductory Electric Circuits.*

DISCUSSION

ac Signal Generator or Function Generator

The exercises you will be doing from this point on include not only the DMM, but also two types of test equipment not used in previous exercises. The first is a *function generator.* The function generator can produce variable amplitude and frequency sinusoidal waveforms. (Most function generators can also be set to produce non-sinusoidal waveforms.) When setting a function generator to produce a specific output waveform, an oscilloscope is used to set the amplitude and frequency of that waveform.

The controls of a signal generator are few. Usually a large dial is used to set the frequency. This dial is used in conjunction with a frequency multiplier selector switch. If you positioned the dial on the number 5 and set the multiplier selector switch to 10^3, then the frequency would be approximately 5000 Hz. (See the basic function generator in Figure 27.1.)

Some of the common controls are listed below:

power switch—turns the generator on.

base dial—sets the significant digit of the frequency.

multiplier selector switch—sets the times ten base multiplier of the frequency.

waveform selector switch—sets the type of waveform needed for the exercise.

output connector—connected with test leads to the circuit under investigation.

dc offset—some generators have this function; references the ac output to a plus or minus dc level.

> Note: Most signal generators have calibration controls that have to be set on "Cal." Look for them on your generator. If they are not set on "Cal," then both the output voltage and the output frequency could be in error.

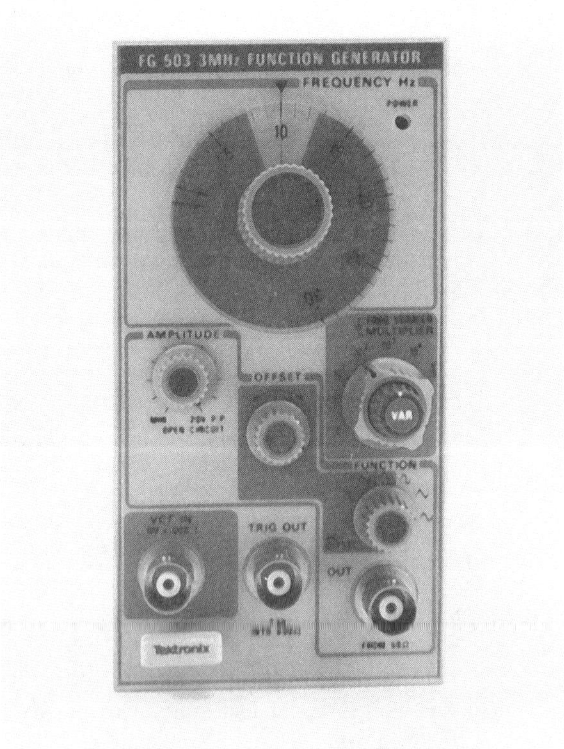

FIGURE 27.1 Basic function generator.

Use the space below to make notes of the differences between your generator and the one described in this exercise.

Oscilloscope

The oscilloscope allows you to observe an electrical signal electronically. This electrical signal could be a dc voltage or an ac voltage waveform, or any other type of electrical waveform. We will use the oscilloscope to determine the frequency and magnitude of a signal; in future exercises, we will use it to determine the phase relationship and frequency responses with other components. The controls of an oscilloscope are a little more involved than those of the audio signal generator, as shown in Figure 27.2. Below are some of the more important controls.

time base control—selects the *time/division* for the sweep circuit; determines the time displacement with each vertical division.

horizontal position control—positions the positive-going waveform to cross the x-axis at a vertical division line.

vertical sensitivity selector—selects the *volts/division* for the y-axis sensitivity, which determines the voltage displacement between each horizontal division.

vertical position adjust—positions the waveform up and down the display screen.

dc, ac, and gnd (common)—allows for the selection of a *dc input* or *ac input*, or grounds the input primarily to position the y-axis sweep line.

trigger selector switch—determines if the waveform triggers on the *positive or negative slope* of the waveform. (+), (−), (auto).

trigger positioning control—determines where on the slope of the waveform the sweep circuit will trigger.

> Note: Most oscilloscopes have calibration controls that have to be set on "CAL." Look for them. They will give false measurements for voltage amplitude as well as frequency if the calibration switches are not set properly.

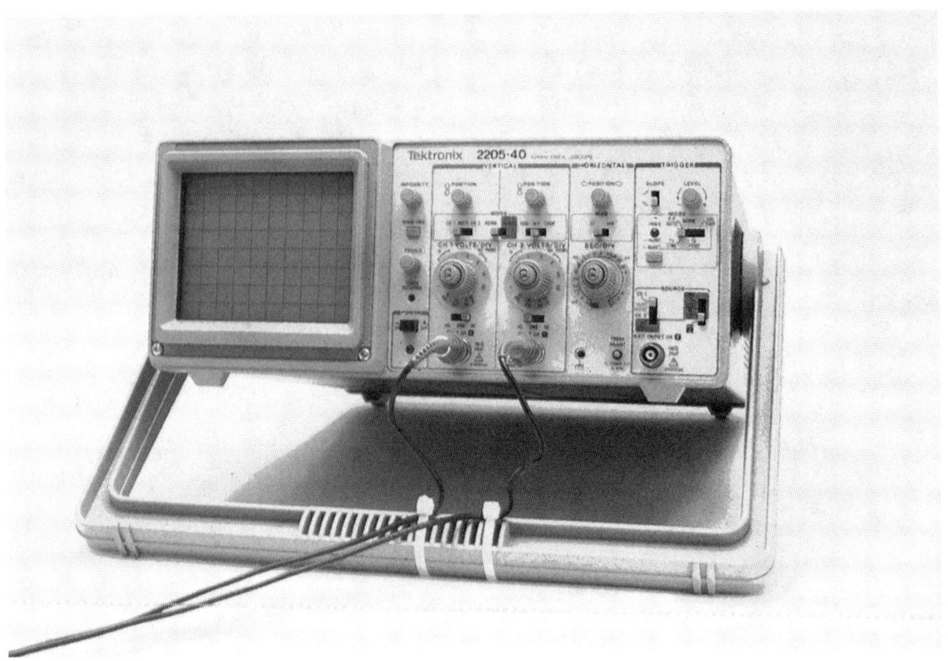

FIGURE 27.2 Basic oscilloscope.

Use the space below to make notes of the differences between your oscilloscope and the one described in this section.

MATERIALS

1 ac signal generator
1 dual-trace oscilloscope
1 DMM
2 D cell batteries and holders
1 test lead

PROCEDURE

dc Measurements

1. Set up the circuit shown in Figure 27.3a. Make sure that the calibration switches on the oscilloscope are set to "Cal".
2. Set the oscilloscope vertical sensitivity to 0.5 volts/division for a ×1 scope probe (or 0.05 volts/division for a ×10 scope probe). Set the time base for 1 msec/div.
3. Center the horizontal trace by selecting the gnd input switch. Then turn the vertical positioning adjustment so that the trace line is centered vertically on the display grid.
4. Use the focus and intensity controls to achieve a narrow, sharp trace across the scope screen.
5. Use the horizontal positioning control to make sure that the trace starts at the left edge of the display screen.
6. Select the dc input switch for Channel A (Ch A).
7. Connect the positive tip of the probe or test lead to node *A*. Connect the negative lead or alligator clip to node *B*.
8. Now measure the deflection of the trace up the display screen. The measurement will be in centimeters of displacement. Record this displacement below.

   Vertical displacement = ______________ cm

   dc voltage = (vertical displacement)(vertical sensitivity)
   = (   )(   ) = ______________

9. Now go through the same procedure for Figure 27.3b. Will there be enough vertical displacement for approximately 6 divisions of displacement? No. Ground the input as before and move the trace down to a minor division *x*-axis so there are at least 6 divisions or more of possible vertical trace sensitivity. You should be able to measure the deflection of the next part.
10. Reset your trace line. Do Steps 7 and 8 to measure the two 1.5 V cells in series as shown in Figure 27.3b. Record your measurement below.

    dc volts = (vertical displacement)(vertical sensitivity)
    = (   )(   ) = ______________

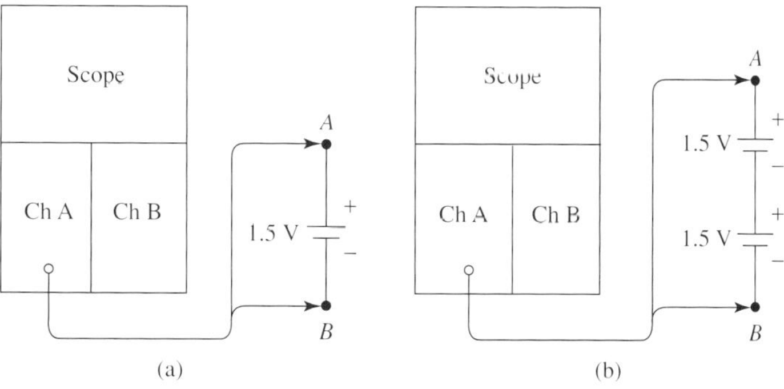

FIGURE 27.3 dc test circuits.

ac Measurements

1. Connect the signal (function) generator to the oscilloscope as shown in Figure 27.4. We will use a frequency of 1000 Hz for our exercise.
2. You must follow several steps each time you use these types of equipment. Set up their controls as follows:

Signal Generator		*Oscilloscope*	
Base dial	10	Input selector	ac
Multiplier	10^{+2}	Time base	0.1 msec/div.
Waveform	Sinusoidal	Vertical sensitivity	1 volt/div.
		Trigger mode	Auto.

3. Make sure that the trace on the scope is centered. You center the trace the same way that you did for dc, with the vertical positioning adjustment. The trace should go completely across the screen. If it does not, you will need to use the horizontal positioning adjustment to make sure that the trace goes from the left edge of the display screen across to the right edge.
4. You need to set the intensity control so that it is not too bright. You will probably need to adjust the focus to make the trace clear and sharp.
5. Make sure the input is in the ac mode. While observing the scope screen, turn up the amplitude control on the signal generator until you have 6 volts ac peak-to-peak.

   How many divisions of vertical deflection did the trace deflect? ________ cm

   (Number of deflections) × (1 volt/div.) = ________ V
6. With the horizontal position adjustment, adjust the waveform so that the positive-going trace comes out of the left side of the screen at the major or center horizontal x-axis.
7. With the base dial of the signal generator, adjust the waveform so it exits on the right side of the screen at the major or center horizontal x-axis. With this accomplished, you have set the controls of the signal generator for an output of 6 V ac at a frequency of 1000 Hz.
8. Now change the volts/div control to 5 volt/div. Change the time base to 10 μs/div, and the multiplier on the generator to 10^3. These changes will give you a frequency of ________ Hz. How many cycles do you have on the screen? ________
9. Increase the voltage output from the generator until you see 20 V peak-to-peak.
10. What do you have to do to the time/div control to see two cycles on the screen?

    ____
11. Switch the volt/div and the time/div to different settings. What happens?

    ____

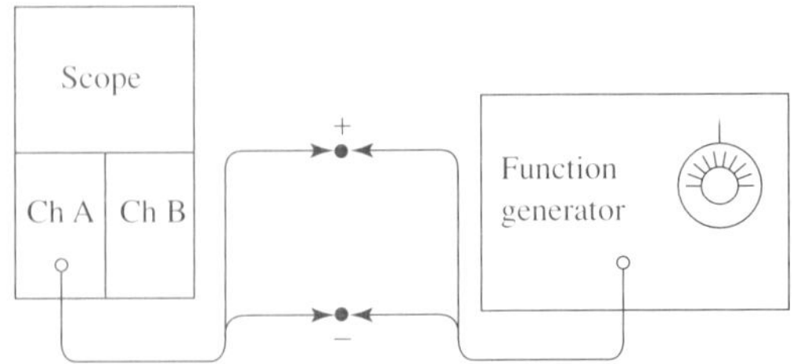

FIGURE 27.4 ac test circuit.

Exercise 28

ac Waveform Measurements

OBJECTIVES

After completing this exercise, you should be able to:

1. Convert frequency to cycle time, and vice versa.
2. Adjust the output from your function generator to a desired frequency.
3. Adjust the output from your function generator to a desired amplitude.

LAB PREPARATION

Review Section 11.1 of *Introductory Electric Circuits*.

DISCUSSION

Frequency and Cycle Time

When investigating the frequency and cycle time of a waveform, we are concerned only with time-related measurements. To measure frequency and cycle time, we need to focus on the x-axis of the oscilloscope screen. The x-axis of the oscilloscope reflects the setting of the time/base control (time/division). We will need three equations for period and frequency measurements:

$$f = \frac{1}{T_C} \qquad T_C = \frac{1}{f}$$

where

$$T_C = \text{cycle time or period of the waveform}$$

$$f = \text{frequency}$$

Frequency and Amplitude

Setting the output amplitude and frequency from your function generator can be accomplished by adjusting the proper generator control while viewing the waveform on the oscilloscope. In this exercise, you will learn the procedure for setting the amplitude and frequency of a sinusoidal output from your function generator.

MATERIALS

1 dual-trace oscilloscope
1 audio signal generator
2 sets of test leads

PROCEDURE

Frequency and Cycle Time

1. Set up the oscilloscope and signal generator as shown in Figure 28.1. Make sure that the polarities of the test leads are connected correctly: positive to positive and negative to negative.
2. Set the frequency adjust control on the signal generator to 8. Set the multiplier to the 10^2 setting. What is the frequency of the generator? ________ Hz
3. Set the oscilloscope's vertical sensitivity to 1 volt/div for the Ch A input and the time/div control to 0.2 msec/div.

> Note: Make sure that you set the trace to zero after you set the volts/div control. Every time you change the volts/div control, you must check the zero position of the trace. Make sure that the oscilloscope and the signal generator are set to their calibrated settings.

4. Adjust the amplitude control on the signal generator until you see at least 5 divisions of displacement peak-to-peak.

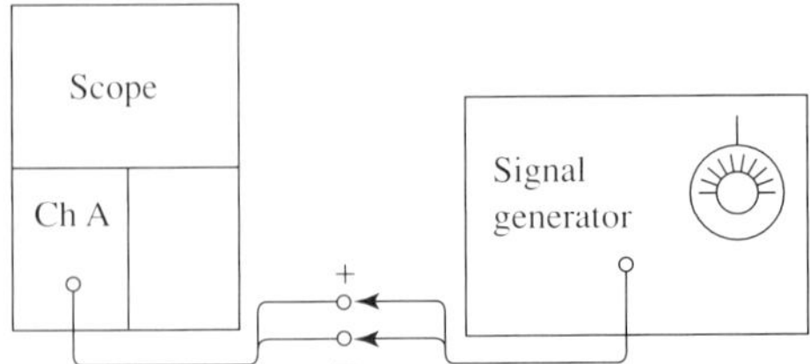

FIGURE 28.1 Oscilloscope and signal generator setup.

5 Set the oscilloscope time base control so that you have at least two complete cycles of the waveform.
6. Reset the oscilloscope time base control to see if you can display 1 to 1½ cycles.

> Note: Make sure that the positive-going trace crosses at the center major x-axis at the left edge of the screen.

7. Now measure the displacement along the x-axis (the number of divisions of displacement it takes to complete one cycle).

   (Displacement ________) × (time base setting ________) = ________

   T_C = ______________

   f = ______________

8. The cycle time of 800 Hz is ______________________.

   Does this cycle time match the T_C of Step 8? ______________________
9. If the cycle time doesn't match, then slowly turn the frequency adjust control on the function generator until the single cycle displacement is 1.25 msec. (This is how you calibrate the frequency of the signal generator that you will need to use for future exercises.)

Frequency and Amplitude

10. Set the function generator to a frequency of 1000 Hz.

    T_C = ______________ msec
11. Set the oscilloscope time base control to 0.1 msec/div.

    Why use the setting of 0.1 msec/div? ______________________

    ____
12. How many cycles do you see on your oscilloscope screen? (You should see only one complete cycle. If not, check with your instructor.)
13. Switch the vertical sensitivity control to 2 volts/div. When we switch the vertical sensitivity control, what happens to the trace? ______________________

    Switch back? ______________________________
14. Once again, make sure you have the signal generator set for 1000 Hz. Use the procedure given in Step 10 above.
15. Increase the amplitude control on the function generator, and watch the oscilloscope until you have a peak-to-peak voltage of 12 V.

You now have an oscilloscope calibrated at 12 V peak-to-peak ac voltage at a frequency of 1000 Hz. If you have a partner, each of you should take a turn and come up with your own voltage and frequency. (The only way that you can learn how to use this equipment is to experiment with different combinations. Try at least four different settings.)

1. Frequency ______________ voltage ______________
2. Frequency ______________ voltage ______________
3. Frequency ______________ voltage ______________
4. Frequency ______________ voltage ______________

Exercise 29

Phase Angle Measurements

OBJECTIVE

After completing this exercise, you should be able to:

1. Determine the phase angle between two waveforms that have the same cycle time but do not reach their peak values or their zero or x-axis crossing at the same time.

LAB PREPARATION

Review Section 11.3 of *Introductory Electric Circuits* on Phase Angle and Measuring Phase Angle.

DISCUSSION

The phase difference between two waveforms is referred to as their phase angle. Two waveforms are in phase when there is no phase difference between them. If two waveforms are out of phase, you can determine the phase angle between them by measuring the distance, in time, between the two waveforms. To do this, you measure the distance, in time, between the points where their positive-going slopes cross the zero or x-axis reference line. In Figure 29.1, the voltage waveform crosses the zero line before the current waveform. There is a phase difference between the two waveforms. The phase angle difference between the current and voltage in a pure inductive or capacitive circuit is 90°, as will be demonstrated in future exercises. In this exercise, your instructor will give you an inductor

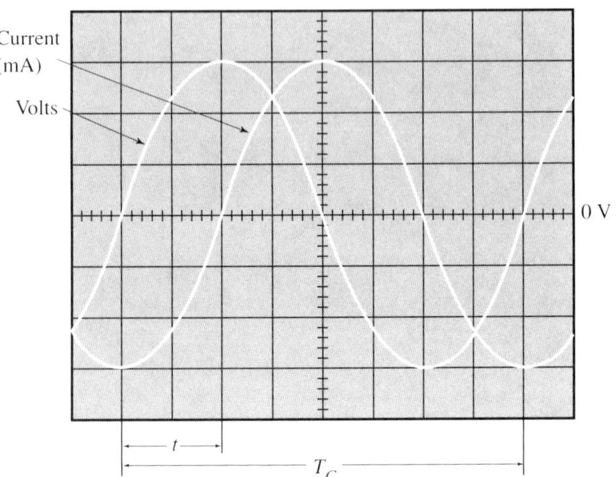

FIGURE 29.1 Phase angle nomenclature.

and a resistor to produce a measurable phase difference between the source voltage and the circuit current.

$$T_C = \text{number of divisions} \times \frac{\text{time}}{\text{divisions}} \qquad t = \text{distance between waveforms} \times \frac{\text{time}}{\text{divisions}}$$

$$\theta = (360°) \times \frac{t}{T_C} \qquad I_S = \frac{V_R}{R}$$

MATERIALS

1 dual-trace oscilloscope
1 signal generator
1 protoboard
1 10 mH inductor
1 1.2 kΩ, 0.5 W resistor
2 sets of test leads

PROCEDURE

1. Measure the ohmic value of R. $R =$ ______________________
2. Set up the circuit shown in Figure 29.2. Make sure that the Ch A positive test lead of the oscilloscope goes to node A and that the positive lead of Ch B of the oscilloscope goes to node B. The negative test leads of Ch A and Ch B go to node C.
   Ch A: volts/div. = 0.5 V
   Ch B: volts/div. = 0.5 V

> Note: Channel A will be the reference channel. Channel A will monitor the source voltage, and channel B will determine the magnitude of the circuit current and its phase shift in relationship to the source voltage. Make sure that both traces are set to zero and all the "Cal" positions are in the correct calibration positions for both the oscilloscope and the signal or function generator.

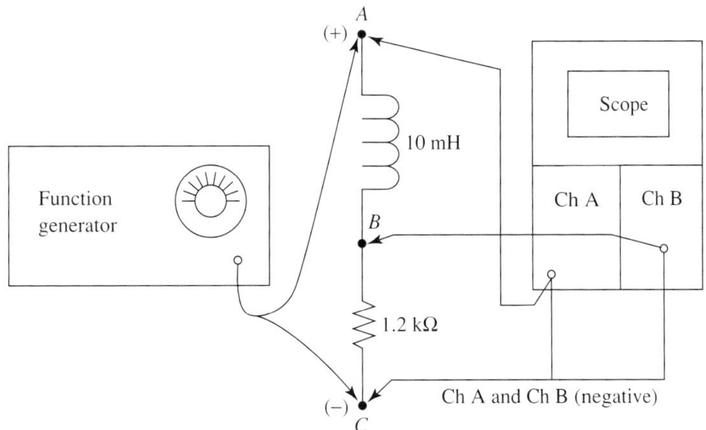

FIGURE 29.2 Phase angle measurements.

3. Set up channel A using the right setting of the trigger sweep control so that you get only one cycle.

> Note: Have channel B in the gnd mode so that you can set up channel A.

4. With both channels set to ac mode, measure the cycle time of the channel A waveform: $T_C =$ __________ μsec.
5. Measure the distance between the two waveforms where they cross the zero-crossing points on the left side of the scope screen: $t =$ __________ μsec.
6. Using the equation above, calculate the phase angle difference between the two waveforms: $\theta =$ __________.

QUESTIONS

1. Two or more waveforms of equal frequency that do not peak at the same time are said to be ______________________________.
2. Any phase difference between two waveforms of equal frequency is called a ______________________________.
3. How can you measure the phase angle between two waveforms of equal frequency? ______________________________

____
4. The phase angle between two or more waveforms is strictly a function of ______________________________.
5. When you change the time base control of an oscilloscope, do you change the waveform? Give reason(s) for your answer. ______________________________

____

Exercise 30

Inductor Characteristics

OBJECTIVES

After completing this exercise, you should be able to:

1. Determine the reactance of an inductor using phase measurements.
2. Compare the nominal and measured values of an inductor.

LAB PREPARATION

Review Sections 13.2 and 13.4 of *Introductory Electric Circuits* on Phase Shift and Inductive Reactance (X_L).

DISCUSSION

As stated in the text, the voltage in a purely inductive circuit leads current by 90°. This relationship is illustrated in Figure 30.1.

A current waveform like the one in Figure 30.1 cannot be displayed directly on an oscilloscope, because oscilloscopes are designed to display *voltages*. However, a waveform that has the same *phase* as the circuit current can be produced using a *current-sensing resistor* (R_S), a low-value resistor that is positioned so that the circuit current passes through it. When current passes through a sensing resistor, a voltage is developed across the resistor that is in phase with that current. Therefore, the voltage across the sensing resistor represents the phase of the circuit current.

Any time resistance is added to a reactive circuit, the phase angles in that circuit are affected. However, if the value of R_S is low enough, the effect on the circuit phase angles

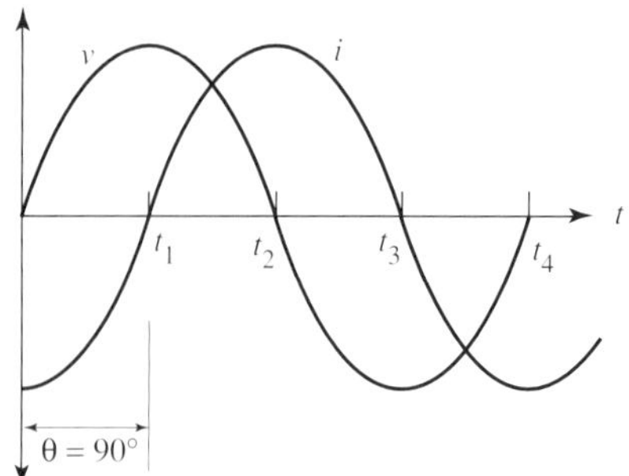

FIGURE 30.1 Phase relationship between inductor voltage and current.

is minimal. Therefore, it is possible to obtain a pair of waveforms that are nearly identical to those in Figure 30.1. Some useful equations for this exercise are given as follows:

$$I_L = \frac{V_S}{R_S} \qquad X_L = 2\pi fL \qquad X_L = \frac{V_{L(\text{rms})}}{I_{L(\text{rms})}}$$

$$\theta = \tan^{-1}\left(\frac{X_L}{R_T}\right)$$

MATERIALS

1 dual-trace oscilloscope
1 signal generator
1 DMM
1 10 mH inductor
1 47 Ω, 0.5 W resistor
1 protoboard
1 set of test leads

PROCEDURE

1. Measure the value of R_S and R_1: R_S = _______, R_1 = _______.

   > Note: R_1 is being used to represent the winding resistance of the coil (R_w).

2. Construct the circuit in Figure 30.2.
   a. Connect channel A to node *A*.
   b. Connect channel B to node *B*.
   c. Connect both channel grounds to node *C*.
3. Using channel A to monitor the output voltage of the signal generator, set the frequency to 10 kHz. Use the oscilloscope to calibrate the frequency of the signal generator. Set the output amplitude of the signal generator to 6 V_{pp} from node *A* to node *C*.
4. With the DMM, measure the voltage drop across the inductor (V_L) and record below.

   $V_{L(\text{rms})}$ = _______
5. With the DMM, measure the voltage drop, V_S, across the sensing resistor, R_S.

   V_S = _______

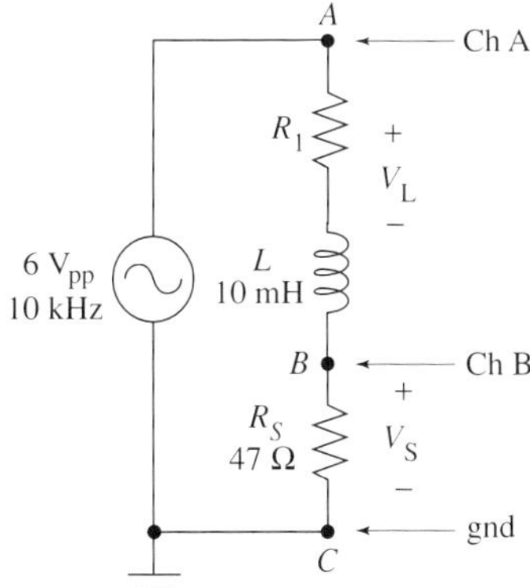

FIGURE 30.2 Inductive test circuit.

6. Predict the current I_L using the measured values of V_S and R_S.

   I_L = _______

7. Predict the value of X_L using the values of V_L and I_L.

   X_L = _______

8. Given the values of X_L and R_T, determine the phase angle θ between the voltage and current.

   θ = _______°

9. Observe the wave shapes on the screen and determine the phase angle between the voltage waveform and the current waveform. The display should be similar to the one in Figure 30.1, with the second waveform crossing the x-axis at the same time as the first waveform is at its peak. How close to 90° is the phase difference between the voltage and the current? _______

10. What would be the percent of difference between the ideal value and the one observed phase difference in Step 9?

    Percent of difference = _______

11. Using the correct equation, determine the value of the inductor in this exercise.

    L = _______

12. What is the percent of difference between the nominal value of the inductor and the measured value of the inductor?

    Percent of difference = _______

13. In your own words, explain the use of the sensing resistor in this exercise.

    ____

    ____

    ____

Exercise 31

Series RL Circuit Characteristics

OBJECTIVES

After completing this exercise, you should be able to:

1. Demonstrate that the ohmic value of resistance is independent of audio frequencies.
2. Demonstrate that the inductive reactance of an inductor increases linearly with increases in frequency.

LAB PREPARATION

Review Sections 13.4 and 14.1 of *Introductory Electric Circuits*.

DISCUSSION

Resistance

You will discover that as you increase the frequency of the source generator in a purely resistive ac circuit with a fixed voltage, the voltage drop across both the circuit resistance and the sensing resistor remain the same. The phase angle remains the same between the voltage and the current.

Inductive Reactance

You will find that as you increase the frequency of the source generator, with the voltage at a fixed value, the inductive reactance of an inductor increases. This relationship will be demonstrated as you observe the voltage drop across an inductor increase as frequency increases, and the voltage drop across a sensing resistor decrease as the frequency increases.

When exploring the parameters of any inductive circuit, you have to take into account the internal resistance of the inductive coil. This small ohmic value can make a big difference in some calculations if you neglect to take it into consideration. For example, if the series resistance is a low value, then you would need to include the internal resistance of the coil. The equation below would need to be used:

$$R_T = R_S + R_w$$

You measure R_w using the ohmmeter function of your DMM.

MATERIALS

1 oscilloscope
1 function generator
1 DMM
1 100 Ω, 0.5 W resistor
1 1 kΩ, 0.5 W resistor
1 10 mH inductor
1 protoboard
1 set of test leads

PROCEDURES

Resistance

TABLE 31.1 Measured Resistance

Resistance	*Measured Values*
$R = 1\ \text{k}\Omega$	
$R_S = 100\ \Omega$	
$R_w =$	

1. Measure all the resistive values and record them in Table 31.1.
2. Construct the circuit in Figure 31.1.
3. Set the function generator at 4 V_{PP} and a frequency of 1 kHz. Record the voltage drops across R and R_S in Table 31.2.
4. Observe the phase shift between the two voltage drops. Record the value of the phase shift in Table 31.2.
5. Increase the frequency as dictated in Table 31.2 and record the appropriate voltage drops and phase shift.

> Note: Make sure that you check the value of the source voltage each time you change the frequency.

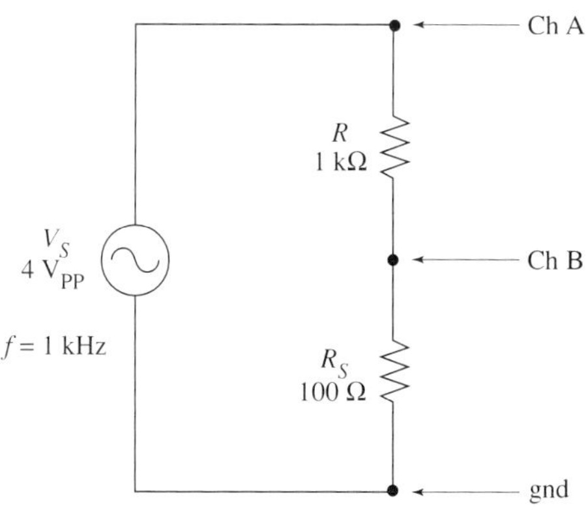

FIGURE 31.1 Resistance test circuit.

TABLE 31.2 Resistive Voltage Drops

Frequency	*V_R (Volts)*	*V_S (Volts)*	*θ (Degrees)*
1 kHz			
2 kHz			
3 kHz			
4 kHz			
5 kHz			

Inductive Reactance

6. Construct the circuit in Figure 31.2.
7. Set the function generator at 4 V_{PP} and the frequency at 1 kHz.
8. Using your DMM, measure the voltage drops across the inductor, V_L and V_S. Record these values in Table 31.3. Measure the phase shift between the source voltage and the current and record this value in Table 31.3.
9. Determine the inductive current (I_L) for each frequency in Table 31.3 using V_S and R_S. Record the value of I_L in Table 31.3.

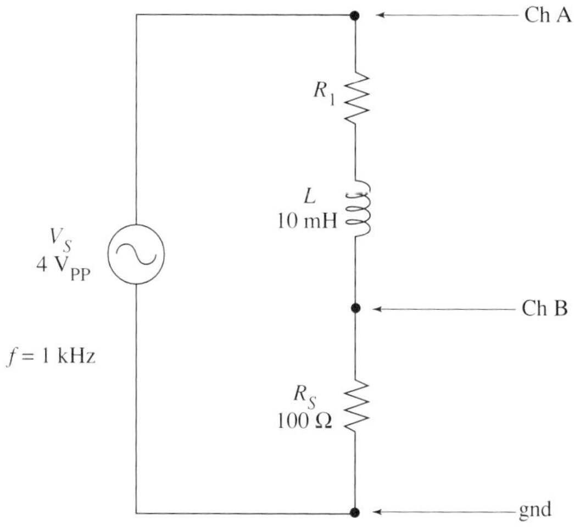

FIGURE 31.2 *RL* test circuit.

TABLE 31.3 Inductive and Resistive Voltage Relationships

Frequency	V_L *(Volts)*	V_S *(Volts)*	θ *(Degrees)*	I_L
1 kHz				
2 kHz				
3 kHz				
4 kHz				
5 kHz				

TABLE 31.4 Inductive Reactance Values

Frequency	*Predicted* X_L	*Measured* X_L	*Predicted* θ	*Measured* θ
1 kHz				
2 kHz				
3 kHz				
4 kHz				
5 kHz				

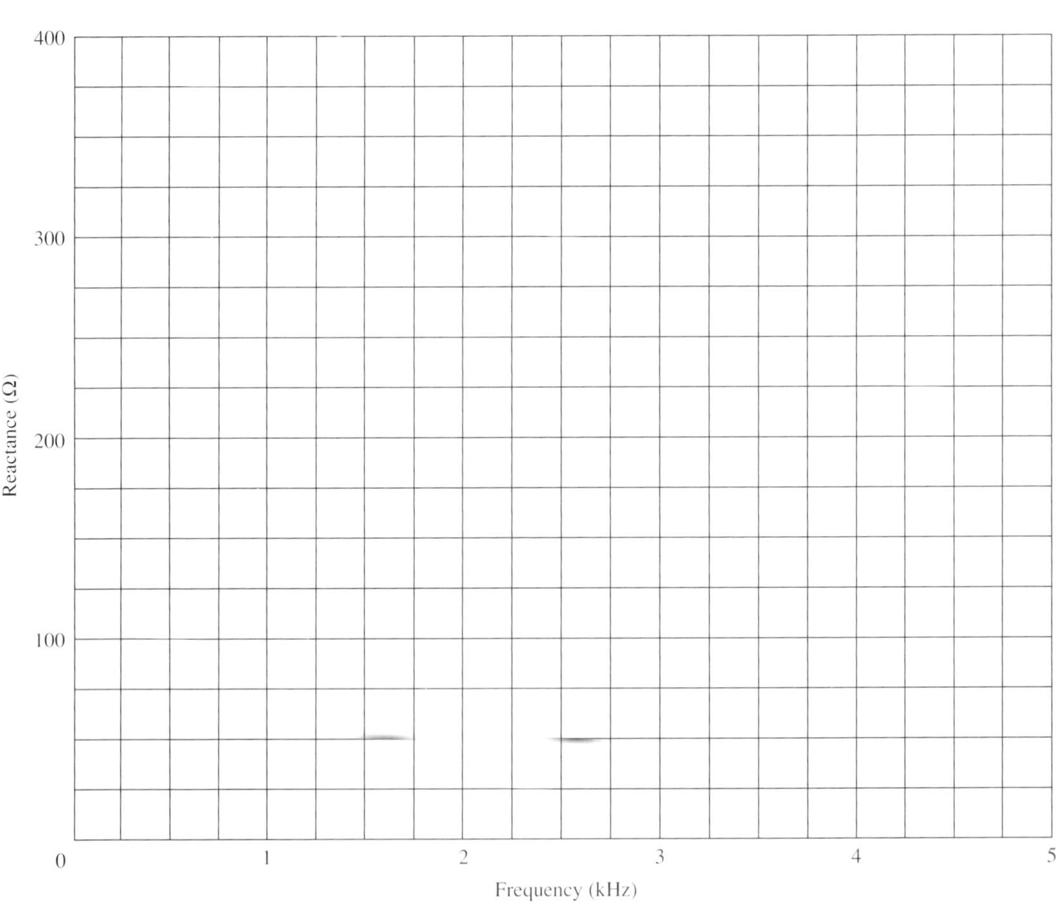

FIGURE 31.3 Inductive reactance versus frequency.

10. Predict the inductive reactance of the inductor for each frequency as indicated in Table 31.4. Predict the phase shift between the voltage drop across the inductor and the current through the inductor. Record these values in the appropriate column.
11. Measure the voltage drop, V_L, and the voltage drop, V_R. With this information, calculate the value of X_L. For each frequency, measure the phase shift between the source voltage and the current. Record this value in Table 31.4.

QUESTIONS

Resistance

1. What did you observe with the phase shifts between the different frequencies of Table 31.2? ____
2. Write a statement about the relationship among the frequency, voltage drop, current, and phase shift for a purely resistive circuit. ____________________
____
____

Inductive Reactance

3. As the frequency increased in Table 31.3, what did you observe about the voltage drop across the inductor? ____
4. As the frequency increased in Table 31.3, what did you observe about the phase shift between the voltage, V_L, and the current I_L? ____________________
____
5. What did you discover about the measured phase shift compared to the predicted value of theta (θ) in Table 31.4? ____________________________________
____
6. When the frequency was set at 4 kHz, was the predicted phase shift equal to the measured phase shift? ____
7. Using the measured values of voltage and current, did the measured values of inductive reactance match the predicted values of Table 31.4? ________________
____
____
8. Write a statement describing the relationship of inductive reactance to an increase in frequency. ____
____
____
9. Take the measured values of X_L from Table 31.4 and plot them on the graph in Figure 31.3. Assume that the inductive reactance is 0 Ω for a frequency of 0 Hz.

Exercise 32

Series RL *Circuit Phase Angles*

OBJECTIVES

After completing this exercise, you should be able to:

1. Apply Kirchhoff's voltage law to the voltage drops across the inductor and the resistor and the source voltage in a series *RL* circuit.
2. Use the oscilloscope to make phase-related voltage readings.

LAB PREPARATIONS

Refer to Section 14.1 of *Introductory Electric Circuits*.

DISCUSSION

In a series *RL* circuit you have to consider the phase relationships between the voltage drops and the source voltage. Kirchhoff's voltage law states that the sum of the voltage drops is equal to the source voltage. If you look only at the magnitudes of the voltage drops across an ac circuit, you will see that the algebraic sum of the voltage drops does not equal the magnitude of the source voltage because of the phase difference between the inductor and resistor voltages. Some of the equations that you will need are given below:

$$X_L = \frac{V_{L(\mathrm{PP})}}{I_{\mathrm{PP}}} \qquad I_{\mathrm{PP}} = \frac{V_{R(\mathrm{PP})}}{R_{(\mathrm{Meas.})}} \qquad Z_T = \frac{V_{\mathrm{PP}}}{I_{\mathrm{PP}}}$$

$$\theta = \tan^{-1}\frac{X_L}{R} \qquad V_S = \sqrt{V_R^2 + V_L^2} \qquad X_L = 2\pi fL$$

MATERIALS

1 oscilloscope
1 function generator
1 1 kΩ, 0.5 W resistor
1 10 mH inductor
1 protoboard
2 sets of test leads

PROCEDURE

1. Measure the resistance of R and R_w. Record these values in Table 32.1.
2. Construct the circuit in Figure 32.1.
3. Set the source for 6 V_{PP} at a frequency of 10 kHz.
4. Measure $V_{R(PP)}$, with respect to V_S, with the oscilloscope. Measure the phase shift between V_S and V_R.

   $V_{R(PP)}$ = ________ T_C = ________

   t = ________ θ_1 = ________
5. Swap the positions of the inductor and the resistor. Measure $V_{L(PP)}$ with respect to V_S and determine phase shift between the two.

   V_L = ________ T_C = ________

   t = ________ θ_2 = ________

   Note: By measuring the two voltages this way, each is compared to the source voltage.
6. Calculate the impedance of the circuit and the current I_{PP} using the appropriate relationships.

   Z_T = ________ I_{PP} = ________

TABLE 32.1 Measured Resistance

Resistance	*Measured Values*
R	
R_w	

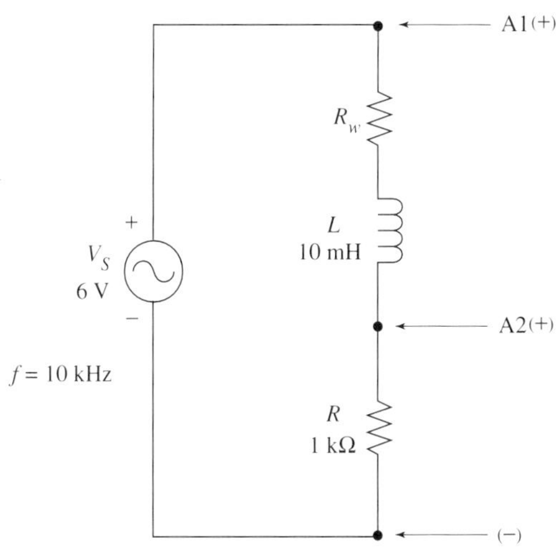

FIGURE 32.1 *RL* characteristics test circuit.

7. Using the Pythagorean theorem, calculate to see if the voltage drops are equal to the source voltage.

   $V_S =$ __________

8. The sum of the two phase angles should equal 90°. Do they?

   $\theta_1 + \theta_2 =$ __________

9. Using a vector diagram, show the relationship between the two voltage drops and the source voltage.

Exercise 33

Series RL Circuit Frequency Response

OBJECTIVES

After completing this exercise, you should be able to:

1. Plot the voltage response due to a frequency change in an *RL* series circuit.
2. Plot the impedance response due to a frequency change in an *RL* series circuit.

LAB PREPARATIONS

Review Section 14.1 of *Introductory Electric Circuits.*

DISCUSSION

For a series ac circuit, the voltage drop across a particular element is directly related to its impedance compared with the other series elements. Since the impedance of an inductor changes with frequency, the voltage drops across the *R* and *L* elements will change with a change in frequency. Use the following equations where needed:

1. $$I_{PP} = \frac{V_{R(PP)}}{R}$$
2. $$Z_T = \frac{V_{S(PP)}}{I_{PP}}$$
3. $$Z_T = \sqrt{R^2 + X_L^2}$$
4. $$\theta = \tan^{-1} \frac{X_L}{R}$$

TABLE 33.1 Measured Resistance

Resistance	*Measured Values*
$R = 330\ \Omega$	
R_w	

MATERIALS

1 oscilloscope
1 function generator
1 330 Ω, 0.5 W resistor
1 10 mH inductor
1 protoboard
2 sets of test leads

PROCEDURE

1. Measure the resistive values listed in Table 33.1 and record them.
2. Construct the circuit of Figure 33.1.

> *Note:* Make sure that all measurements are made with the oscilloscope and the function generator sharing a common ground.

3. Set the source voltage, V_S, to 6 V_{PP} at a frequency of 1 kHz.
4. Measure and record the voltage drops $V_{L(PP)}$ and $V_{R(PP)}$ for all the frequencies listed in Table 33.2.

> *Note:* Verify that V_S remains set for 6 V for each change of frequency.

TABLE 33.2 *RL* Voltage and Current Measurements

Frequency	$V_{L(PP)}$	$V_{R(PP)}$	I_{PP}
1 kHz			
2 kHz			
3 kHz			
4 kHz			
5 kHz			
6 kHz			
7 kHz			
8 kHz			
9 kHz			
10 kHz			

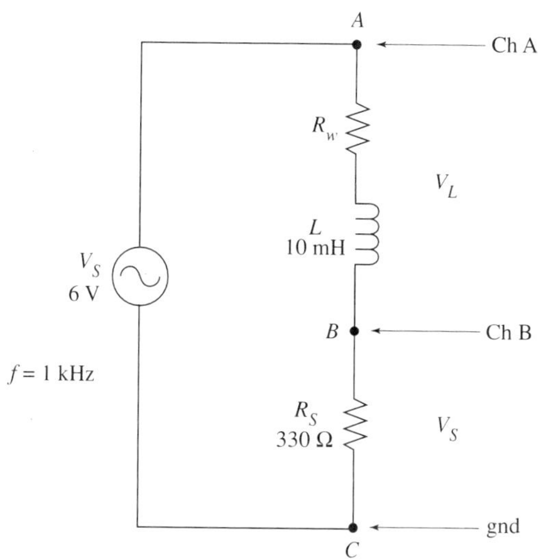

FIGURE 33.1 *RL* test circuit.

5. Calculate the value of I_{PP} for each frequency as indicated in Table 33.2. Record these values in Table 33.2.
6. On the graph in Figure 33.2, plot the curves of $V_{L(PP)}$ versus frequency and $V_{R(PP)}$ versus frequency using the values in Table 33.2.
7. Using the appropriate scale, sketch the curve of I_{PP} on the graph in Figure 33.2.
8. Take the values of I_{PP} from Table 33.2 and record them in the appropriate column in Table 33.3.
9. Using the two methods indicated in Table 33.3, calculate the total impedance for each frequency of the circuit.
10. Using the total impedance from either column in Table 33.3, plot the curve of Z_T on the graph in Figure 33.3. Is this a linear or nonlinear curve? Explain your answer. ______________________________

____

TABLE 33.3 Impedance Versus Frequency

Frequency	$V_{R(PP)}$	I_{PP}	Z_T *(Equation #2)*	Z_T *(Equation #3)*
1 kHz				
2 kHz				
3 kHz				
4 kHz				
5 kHz				
6 kHz				
7 kHz				
8 kHz				
9 kHz				
10 kHz				

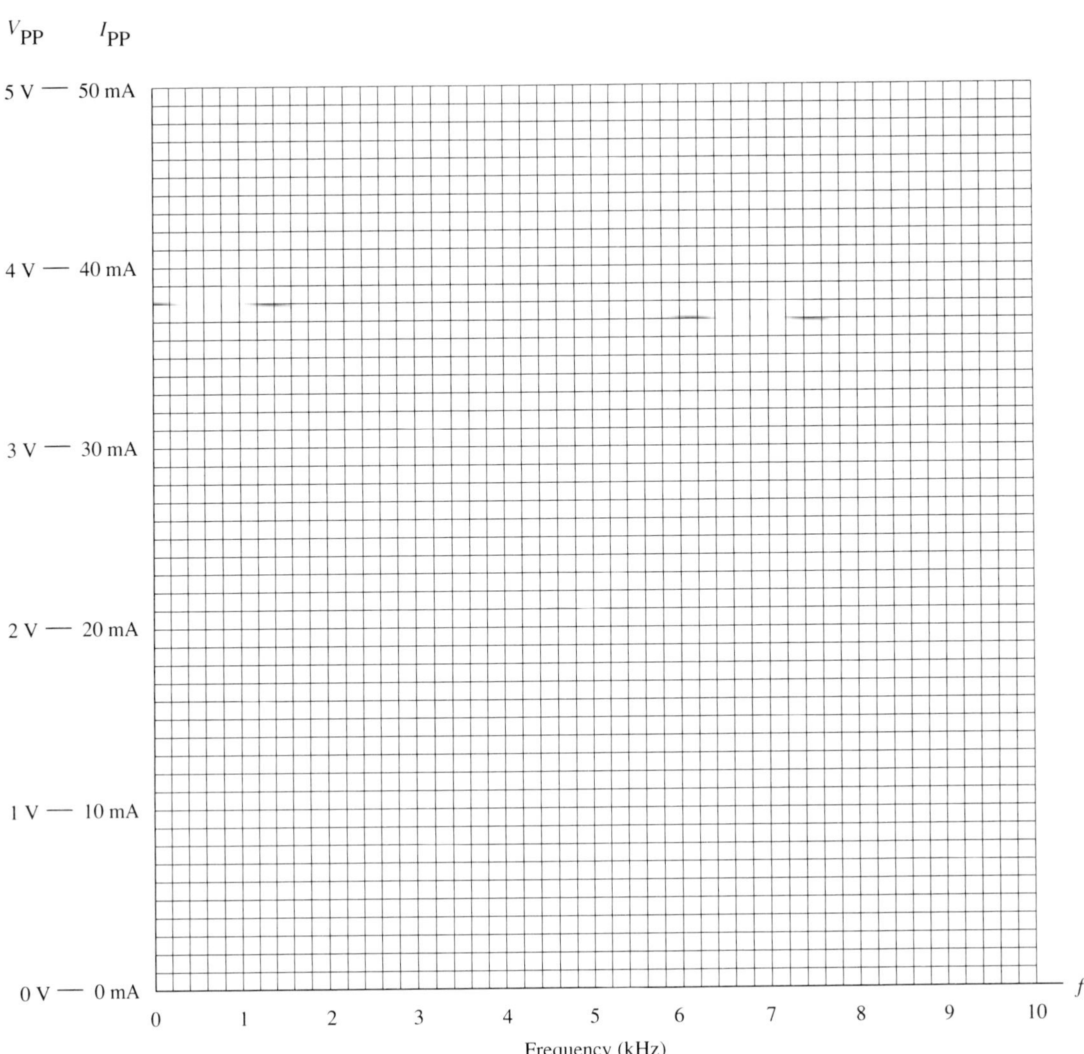

FIGURE 33.2 $V_{L(PP)}$, $V_{S(PP)}$, and I_{PP} curves.

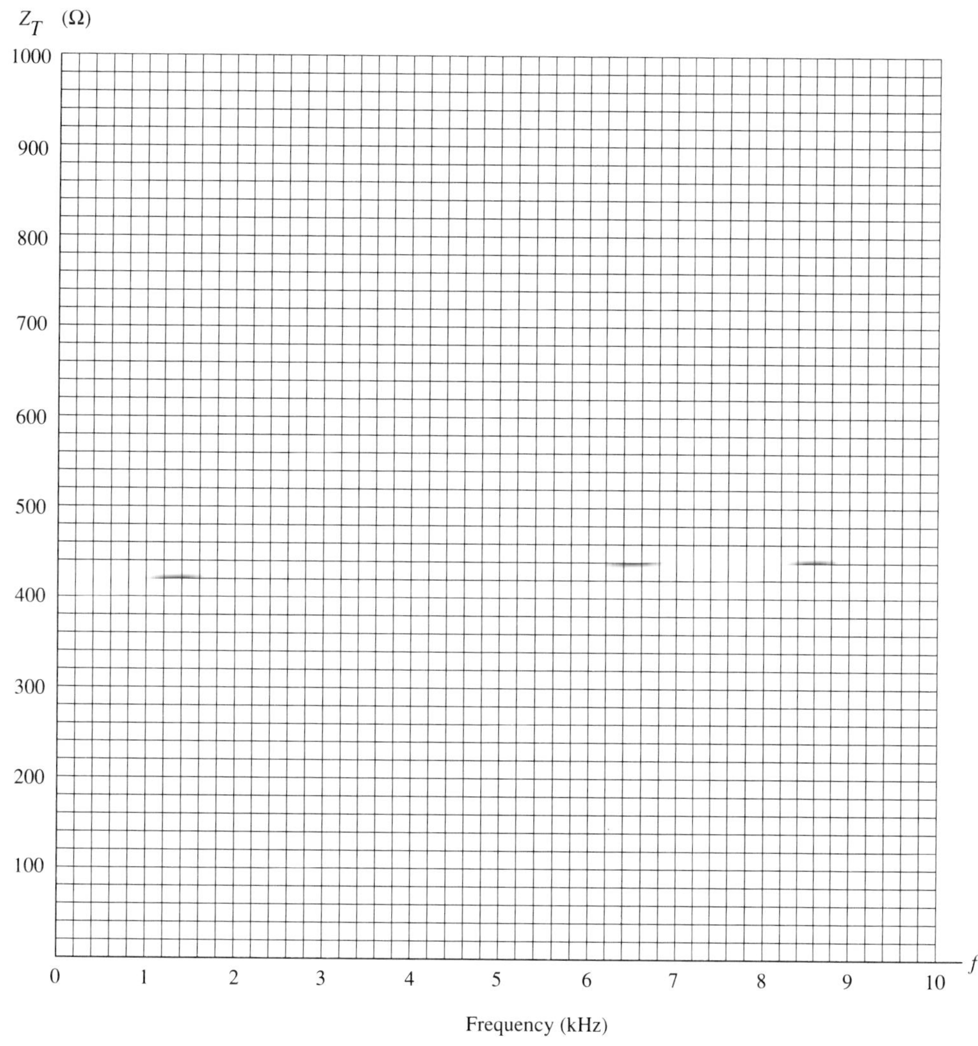

FIGURE 33.3 Impedance versus frequency curves.

QUESTIONS

1. In Step 2, why is it important that the oscilloscope and the function generator share a common ground? ____

____

____

2. Why is it important that you check the source voltage each time you change the frequency? ____

____

3. Does the curve of I_{PP} follow the curve of $V_{R(PP)}$? Why? ________________

____

____

Exercise 34

Parallel RL *Circuit Frequency Response*

OBJECTIVE

After completing this exercise, you should be able to:

1. Perform a complete frequency response analysis of an *RL* parallel circuit.

LAB PREPARATION

Review Section 14.3 of *Introductory Electric Circuits*.

DISCUSSION

In previous exercises, you found that Kirchhoff's voltage law holds true for series ac circuits. In this exercise you will find that Kirchhoff's current law holds true for ac parallel circuits. Kirchhoff's current law states that the sum of the phasor currents entering and leaving a node is equal to zero. All the pertinent equations that you will need for this exercise are given in Section 14.3 of the textbook.

MATERIALS

1 oscilloscope
1 function generator
1 DMM

TABLE 34.1 Resistive Measurements

Resistance	*Measured Values*
$R_S = 10\ \Omega$	
$R = 470\ \Omega$	
R_w	

2 10 Ω, 0.5 W resistor
1 470 Ω, 0.5 W resistor
1 10 mH inductor
1 protoboard
2 sets of test wires

PROCEDURE

1. Measure and record all the resistive values in Table 34.1.
2. Predict the values listed in Table 34.2. Show your work.
3. Construct the circuit in Figure 34.1. Set the frequency generator to 6 V_{PP}, and the frequency to 2 kHz.
4. Measure the current $I_{L(PP)}$ in the inductive branch of the test circuit. Use the 10 Ω sensing resistor in series with the inductor to determine I_L from the voltage drop across the resistor. Record this measured value of I_L in Table 34.2.
5. With the measured value of $I_{L(PP)}$, compute the measured value of X_L. Record this value in Table 34.2.

TABLE 34.2 Predicted and Measured Values of Circuit Variables

Variables	*Predicted Values at 2 kHz*	*Measured Values at 2 kHz*	*Predicted Values at 10 kHz*	*Measured Values at 10 kHz*	*Effect on Variable*
X_L					
$I_{L(PP)}$					
$I_{R(PP)}$					
$I_{T(PP)}$					
θ					
Z_T					
P_X					
P_R					
$P_{(APP)}$					

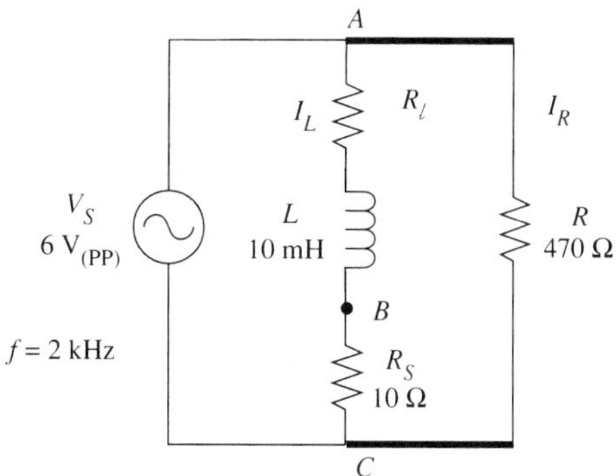

FIGURE 34.1 Parallel *RL* test circuit #1.

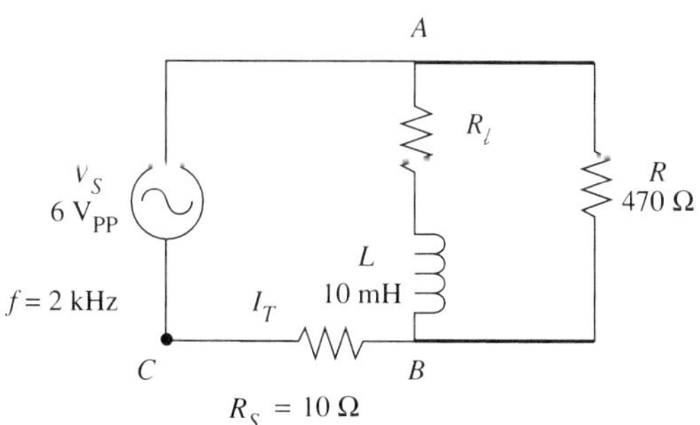

FIGURE 34.2 Parallel *RL* test circuit #2.

6. Measure the voltage, $V_{R(\text{rms})}$, with the DMM. Calculate the measured value of the current, I_R. Record this value in Table 34.2.

   $I_{R(\text{rms})} = I_{R(\text{PP})} =$ ________ mA.

   > Note: The DMM readings are rms values. You need to convert the rms readings into peak-to-peak values.

7. Measure the total circuit current, $I_{T(\text{PP})}$, using the modified test circuit shown in Figure 34.2. Record this value in Table 34.2.
8. Using the two currents, I_L and I_R, calculate the measured value of the phase shift, θ, between the source voltage waveform, V_S, and the total circuit current, I_S. Record this angle in Table 34.2.
9. Using the measured values of V_S and the current I_T, compute the measured value of the circuit impedance, Z_T. Record this value in Table 34.2.
10. Using the measured values of I_L and I_R, determine the reactive power and the resistive power. Record these values in Table 34.2.
11. Using the measured values of P_X and P_R, determine the apparent power, $P_{(\text{APP})}$. Record this value in Table 34.2.
12. Repeat Steps 3 through 11 with the frequency set to 10 kHz. The source voltage will still be 6 V_{PP}.

QUESTIONS

1. Using the source voltage $V_S = 6 \text{ V} \angle 0°$ as your reference angle, sketch the vector angles of I_R and I_L, and determine the vector magnitude and angle of I_S. Use the grid in Figure 34.3 for your sketch. Use $f = 2$ kHz.

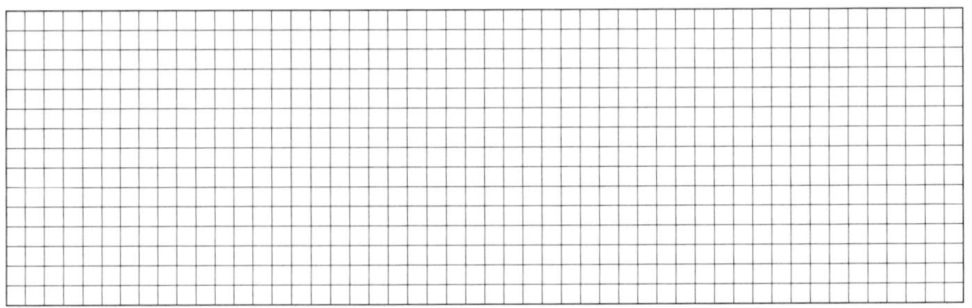

FIGURE 34.3 Phasor diagram of the circuit currents.

2. Does the geometric sum of the currents I_R and I_L equal the magnitude of the source current I_S? ____________________
3. In your own words, state Kirchhoff's current law from your results.

____

____

____

Exercise 35

RL *Transient Response*

OBJECTIVES

After completing this exercise, you should be able to:

1. Determine the steady state response of an *RL* circuit to a dc voltage source.
2. Examine the transient behavior of an *RL* circuit in terms of its voltage and current responses due to a square wave input voltage source.

LAB PREPARATION

Review Section 14.5 of *Introductory Electric Circuits.*

DISCUSSION

A circuit where the input voltage goes from one dc voltage level to another in an instant is called a switching circuit. For resistive circuits, the response for the voltage drops, and current is instantaneous and linear. With an *RL* circuit, the inductor opposes any change in the circuit current, as discussed in Section 13.1 in the textbook about Lenz's law. The current rises to its maximum value over time, as shown in Figure 35.1.

The waveforms for V_R and I are what you would expect, where the voltage drop and the current track each other. When t_1 = zero, there is zero voltage drop and zero circuit current for the resistor. Therefore, the total voltage (source voltage) is dropped across the inductor at t_0 = zero seconds, or when the source is switched on to its positive pulse width. Kirchhoff's voltage law states that the sum of the voltage drops is equal to the source

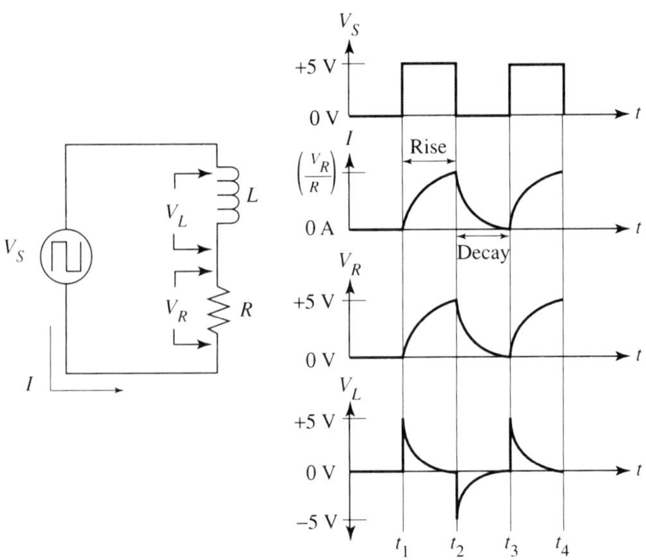

FIGURE 35.1 *RL* circuit waveforms.

voltage. From Figure 35.1, you can see that the sum of the voltage drops is equal to zero. You can also see that as time increases, the sum of the voltage drops V_R and V_L continues to equal zero. At a specified time, you see that the voltage, V_R, goes to the value of the source voltage and the voltage, V_L, goes to zero.

First, you will look at the dc steady state condition of an *RL* circuit. Next, you will investigate the transient response, or the time it takes for the voltage, V_R, to equal the source voltage and V_L to go to approximately zero volts. When this happens, the circuit has reached its steady state condition. The following equations should help you understand the exercise.

Current

$$\text{Rise: } \frac{I_t}{I_{pk}} = 1 - e^{-x} \qquad I_t = I_{pk}\,(1 - e^{-x})$$

$$\text{Decay: } \frac{I_t}{I_{pk}} = e^{-x} \qquad I_t = I_{pk}\,e^{-x}$$

Voltage

$$\text{Rise: } \frac{V_t}{V_{pk}} = 1 - e^{-x} \qquad V_t = V_{pk}\,(1 - e^{-x})$$

$$\text{Decay: } \frac{V_t}{V_{pk}} = e^{-x} \qquad V_t = V_{pk}\,e^{-x}$$

$$t = \text{time} \qquad x = \frac{t}{\tau} \qquad \tau = \frac{L}{R}$$

MATERIALS

1 oscilloscope
1 function generator

1 DMM
1 dc power source
1 1 kΩ, 0.5 W resistor
1 10 mH inductor
1 SPST switch
1 protoboard
2 sets of test leads

PROCEDURE

dc Steady State

TABLE 35.1 Measured Resistances

Resistance	*Measured Values*
1 kΩ	
R_w	

1. Measure and record the resistances in Table 35.1.
2. With the switch open, construct the circuit as shown in Figure 35.2.
3. Predict the values as shown in Table 35.2.
4. Close the switch and, with your DMM, measure the voltage drop across V_L and V_R. With the measured value of V_R, calculate the value of I. Record these values in Table 35.2.
5. Construct the circuit shown in Figure 35.3.
6. Set up the function generator to give you a 10 kHz square wave output. Set the output voltage to give you a 10 V_{PP}.
7. Set up the oscilloscope:

Time base	10 μsec/div.
Ch A and Ch B inputs	dc
Sensitivity	Ch A: 1 V/div.
	Ch B: 0.5 V/div.

> *Note:* By using the dc input mode, you assume that the bottom of the square wave is the zero reference point from which you will be making your measurements.

TABLE 35.2 Steady State Circuit Values

Components	*Predicted Values*	*Measured Values*
V_L		
V_R		
I		

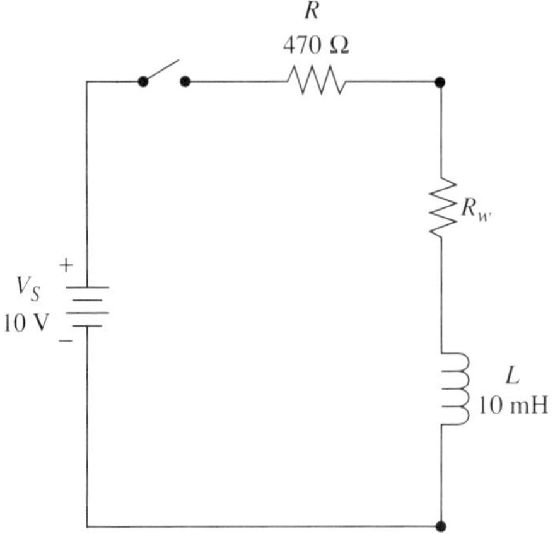

FIGURE 35.2 *RL* steady state test circuit.

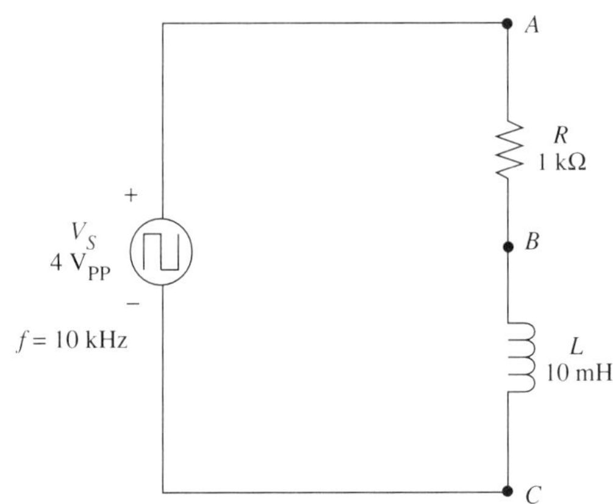

FIGURE 35.3 Transient state *RL* test circuit.

8. Determine the values of V_L, V_R, and I for the first five time constants of the input waveform. Record these values in Table 35.3.
9. Using the time base control, adjust the display so that a single pulse is spread across 5 major divisions on the oscilloscope display grid. This can be accomplished by either:
   a. setting the scope time base as described above, then adjusting the time base "CAL" control to obtain a pulse that is 5 divisions wide. OR
   b. setting the scope time base as described above, then adjusting the function generator frequency control to obtain a pulse that is 5 divisions wide.

   The full 5 time constants of the pulse width are shown in Figure 35.4.
10. With the pulse width display being 5 major divisions wide, the time from each major division to the next represents one time constant. From the display, determine the values of V_L and V_R at the end of each time constant. Record these values in Table 35.4.

TABLE 35.3 Predicted Transient Values

Circuit Predictions	*0τ*	*1τ*	*2τ*	*3τ*	*4τ*	*5τ*
V_L						
V_R						
I						

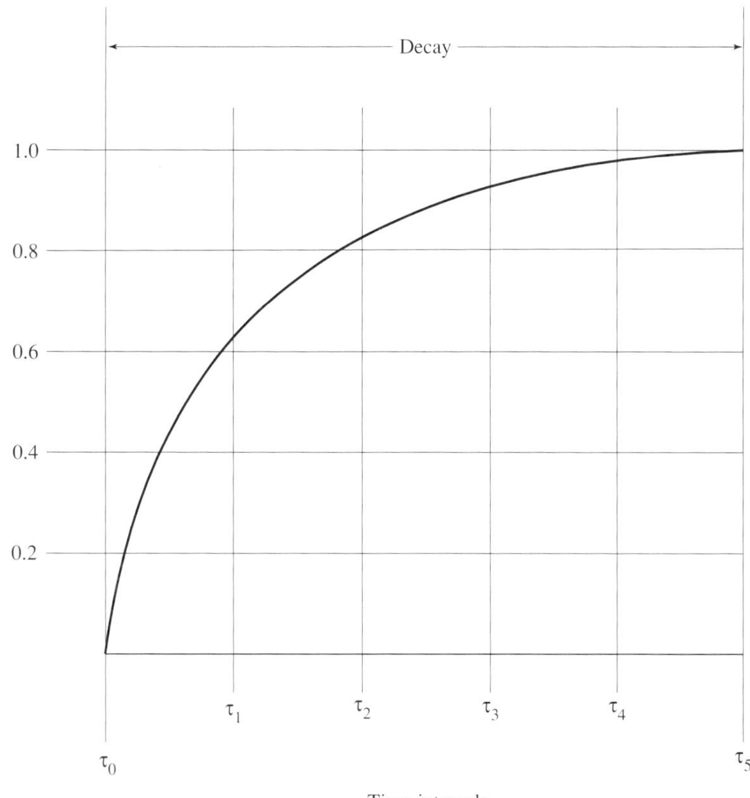

FIGURE 35.4 The decay portion of the universal curve.

TABLE 35.4 Measured Transient Values

Circuit Measurements	0τ	1τ	2τ	3τ	4τ	5τ
V_L						
V_R						
I						

Note: First measure V_L with the circuit as shown. Then you will have to interchange the inductor and resistor and take your measurements of V_R.

11. From the measured values of V_R, calculate the values of I for each time constant. Record these values in Table 35.4.

QUESTIONS

1. Refer to Table 35.2. Are the values of V_R and V_L what you expected? Explain your answer. ____

____

____

2. What is the value of the time constant tau (τ) for the circuit of Figure 35.3?

   τ = ____________________

3. How long would it take for the circuit of Figure 35.3 to reach its steady state condition? Show your work. ____________________

4. Refer to Table 35.4. Graph the curves of the inductor voltage V_L and the voltage drop of V_R on the graph in Figure 35.5. Choose the appropriate scales and labeling for the voltages and the time t. Graph the five time constants.

5. At the time where $t = 25$ μsec, what is the value of V_L from the graph?

   V_L = ____________________

6. Predict what the voltage V_R will be when $t = 25$ μsec.

   ____

7. Are the predicted values of Table 35.3 approximately equal to the values of Table 35.4? Explain your answer in terms of percent of error in V_L.

   ____

   ____

   ____

   ____

   ____

   1τ = __________ % 4τ = __________ %

   2τ = __________ % 5τ = __________ %

   3τ = __________ %

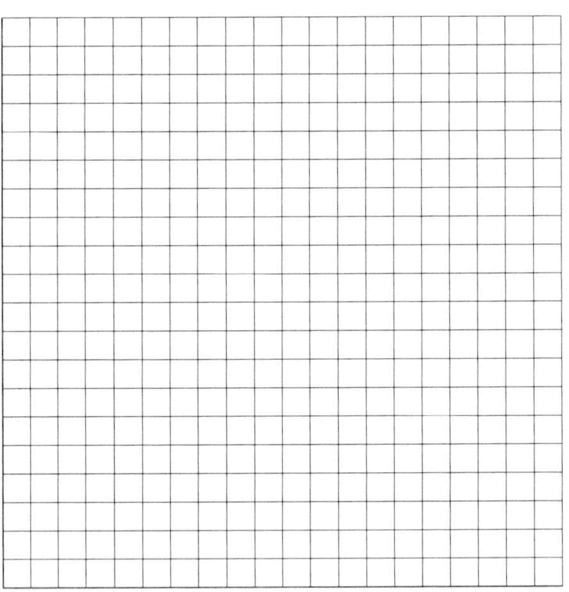

FIGURE 35.5 Time constant voltages V_L and V_R.

Exercise 36

Transformers

OBJECTIVES

After completing this exercise, you should be able to:

1. Determine the phase relationship between the input signal and the output signal of a transformer.
2. Determine the voltage and current relationship of a transformer, its turns ratio, and whether it is a step-up or a step-down transformer.
3. Determine the relationship between the power generated in the primary and the power dissipated in the secondary.

LAB PREPARATION

Review Sections 15.1, 15.2, and 15.3 of *Introductory Electric Circuits*.

DISCUSSION

General

A transformer is a device that transfers energy from its input *(primary)* winding, by electromagnetic induction, to its output *(secondary)* winding(s). Whenever there is a change in primary voltage, the component generates a changing secondary voltage. The phase relationship between the primary and secondary voltages is assumed to be 0° or 180°, depending on the orientation of the primary and secondary windings.

Step-Up and Step-Down Voltage Relationships

The secondary voltage of a transformer may be greater than, equal to, or less than the primary voltage. The relationship between the magnitudes of the input and output voltages is determined by the *turns ratio* of the component; that is, the ratio of primary turns to secondary turns. The relationship among the primary voltage, secondary voltage, and turns ratio is given as

$$\frac{V_P}{V_S} = \frac{N_P}{N_S}$$

Current and Power

The current of the primary is related to the secondary by the following relationship:

$$V_P \times I_P = V_S \times I_S$$

For any transformer, this equation indicates that the power generated by the primary is (ideally) equal to the power dissipated by the secondary. The current of the primary is related to the current of the secondary in the following equation.

$$\frac{I_P}{I_S} = \frac{N_S}{N_P}$$

MATERIALS

1 oscilloscope
1 function generator
1 DMM
1 10 Ω, 0.5 W resistor
1 100 Ω, 0.5 W resistor
1 filament transformer—120 V primary, 12.6 V secondary
3 sets of test leads
1 protoboard

PROCEDURE

Phase Relationship

1. Measure and record the resistor values listed in Table 36.1.
2. Construct the circuit in Figure 36.1.

> Check: Make sure that the calibration controls are set to "Cal." on both the function generator and the oscilloscope as you set up this exercise.

3. Set the function generator for 10 V_{PP} at a frequency of 60 Hz. Connect Ch A across the primary of the transformer and Ch B across the secondary of the transformer, as shown in Figure 36.1.

TABLE 36.1 Measured Resistance

Resistance	*Measured Values*
10 Ω	
100 Ω	

4. Set the controls of the oscilloscope to display at least one complete cycle of the input signal.
5. Observe the waveshapes of both the input signal and output signal of the transformer. Are they in phase? Explain your answer. ____________________

____

Turns Ratio

6. With the input set at 10 V_{PP}, what is the measured output value of the voltage? V_S = ________ V_{PP}
7. Determine the turns ratio of your transformer.

$$\frac{V_P}{V_S} = \frac{N_P}{N_S} = \_\_\_\_ : \_\_\_\_$$

8. Using the above turns ratio, calculate the magnitude of the output voltage. V_S = ________ V_{PP}

Voltage, Current, and Power

9. Construct the circuit of Figure 36.2.
10. Set the voltage at 10 V_{PP} across the primary. Set the frequency of the generator at 60 Hz. Make sure that you have at least one complete cycle of the test signal.
11. Measure the voltage across the primary resistor. Measure the voltage across the load resistor. Record these values in Table 36.2.

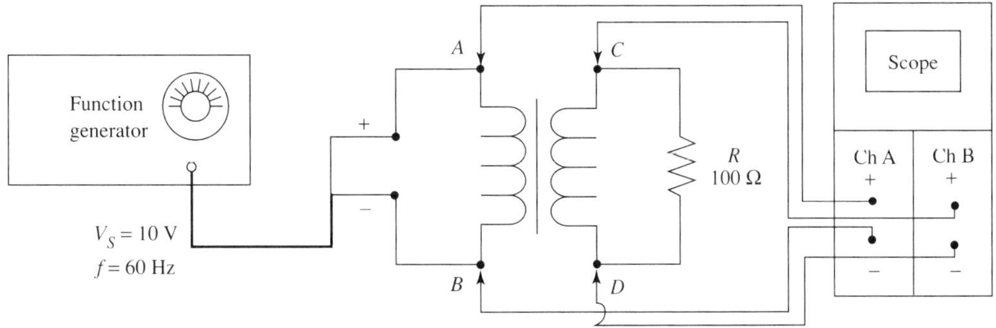

FIGURE 36.1 Phase and turns ratio relationships.

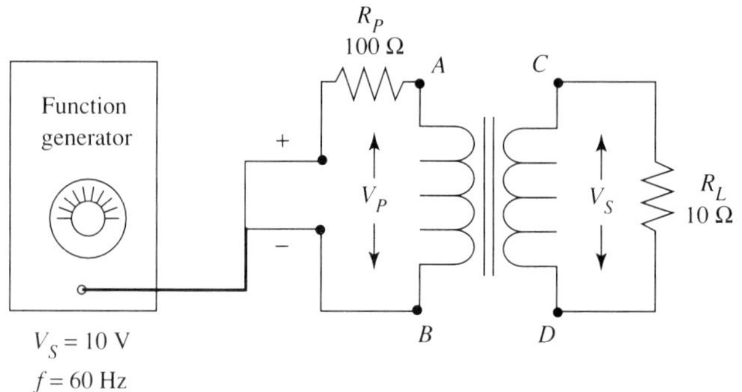

FIGURE 36.2 Test circuit for volts, current, and power.

TABLE 36.2 Voltage, Current, and Power Measurements

R_L	V_P	$V_S = V_L$	I_P	I_S	P_{PRI}	P_{SEC}
10 Ω						

12. Now measure the primary circuit current and the secondary circuit current

    $I_P =$ ______________ $I_S =$ ______________

13. Using the appropriate voltage and current readings, calculate the power generated by the primary circuit and the power dissipated by the secondary. Record these values in Table 36.2.

QUESTIONS

1. What would happen to your voltage reading in Step 7 if you hadn't used your DMM? ______________________________

   ____

2. Did the voltage drop across the primary windings of Step 11 and the voltage drop across the secondary compare to the turns ratio of Step 7 ? ______________

   ____

3. For the 10 Ω load resistor, did the power generated by the primary circuit equal the power dissipated by the secondary circuit from the recorded values of Table 36.2? ______________________________

   Percent of error ________

Exercise 37

Capacitor Characteristics

OBJECTIVES

After completing this exercise, you should be able to:

1. Determine the capacitance of a capacitor.
2. Compare the nameplate and measured values of the capacitor capacitance.

LAB PREPARATION

Review Section 16.4 of *Introductory Electric Circuits* on Capacitance Reactance (X_C).

DISCUSSION

When you did Exercise 30 on Inductor Characteristics, you discovered that the voltage led the current by 90°. For a purely capacitive circuit, the current leads the voltage by 90°, as shown in Figure 37.1. As in Exercise 30, you will use a sensing resistor to measure the current, I_C, in the series capacitive circuit. Some of the equations you may need are given below.

$$I_C = \frac{V_{RS}}{R_S} \qquad X_C = \frac{1}{2\pi f C} \qquad C = \frac{1}{2\pi f X_C} \qquad \theta = \tan^{-1}\frac{R}{-X_C}$$

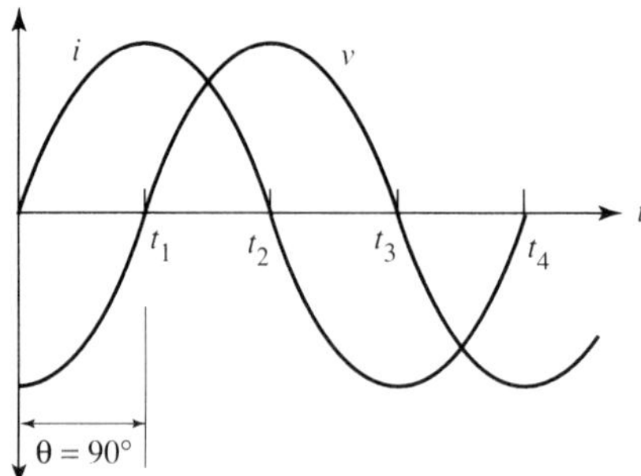

FIGURE 37.1 Phase relationship between capacitor voltage and current.

MATERIALS

1 oscilloscope
1 DMM
1 function generator
1 47 Ω, 0.5 W resistor
1 10 μf capacitor
1 protoboard
2 sets of test leads

PROCEDURE

1. Measure the value of R_S. R_S = __________
2. Construct the circuit in Figure 37.2.
3. Set the function generator for a 6 V_{PP} sine wave at a frequency of 1 kHz.
4. With the DMM, measure the voltage across the sensing resistor (R_S) and record this value below.

   $V_{RS(\text{rms})}$ = __________
5. With the DMM, measure the voltage across the capacitor (V_C) and record this value below.

   $V_{C(\text{rms})}$ = __________
6. Calculate the value of I_C using the measured values of V_{RS} and R_S.

   $I_{C(\text{rms})}$ = __________
7. Calculate the value of X_C using the rated value of the capacitor and the circuit operating frequency.

   X_C = __________

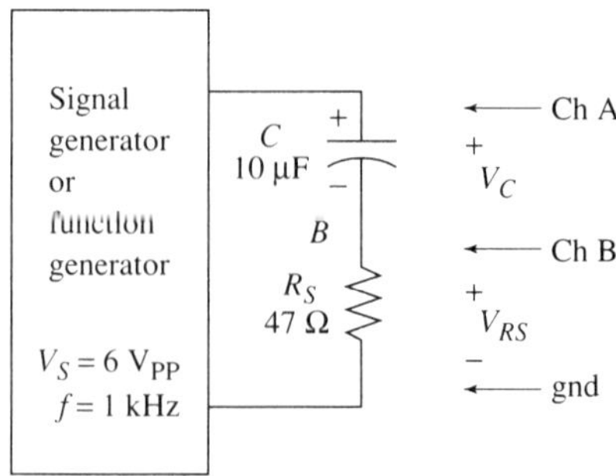

FIGURE 37.2 Capacitive test circuit.

8. Calculate the value of X_C using the measured values of V_{RS} and I_C.

   X_C = __________

9. Determine the capacitor value in μf.

   C = ______________

10. Using the values of X_C (from step 8) and R, determine the phase angle between the voltage and the current.

    θ = __________

11. Observe the waveshapes on the oscilloscope screen and determine the phase angle between the voltage waveform and the current waveform. (It should look like Figure 37.1, with the voltage waveform crossing the x-axis at the same time as the current waveform is at its peak.) How close is it to the 90° phase difference between the current and the voltage? __________

12. What would be the percent of difference between the calculated value of θ in Step 10 and the observed phase difference in Step 11?

    Percent of difference = __________

13. What is the percent of difference between the nominal value of the capacitor and the determined value of the capacitor?

    Percent of difference = __________

Exercise 38

Capacitive Reactance

OBJECTIVE

After completing this exercise, you should be able to:

1. Demonstrate that the ohmic value of capacitive reactance decreases from an open circuit, or infinite ohms, inversely almost to zero ohms geometrically as frequency increases.

LAB PREPARATION

Review Section 16.4 of *Introductory Electric Circuits* on Capacitive Reactance.

DISCUSSION

As stated in the text, capacitive reactance varies inversely with operating frequency. This relationship can be demonstrated using the following equations:

$$X_C = \frac{1}{2\pi f C} \qquad I_R = I_C = \frac{V_{RS}}{R_S} \qquad X_C = \frac{V_{C(\text{rms})}}{I_{C(\text{rms})}}$$

In this exercise, you will again be using a sensing resistor to determine the series current.

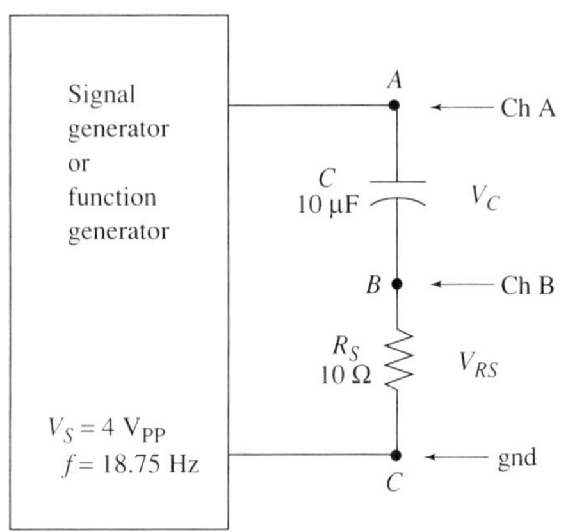

FIGURE 38.1 Capacitive reactance test circuit.

MATERIALS

1 oscilloscope
1 function generator
1 DMM
1 10 Ω, 0.5 W resistor
1 10 μf capacitor
1 protoboard
2 sets of test leads

PROCEDURE

1. Measure the resistor value R_S. R_S = ______________________
2. Construct the circuit of Figure 38.1.
3. Predict the capacitive reactance for the frequencies listed in Table 38.1. Record these values in Table 38.1.
4. Set the function generator for 4 V_{PP} at the first frequency of Table 38.1.
5. With the DMM, measure the voltage drop, $V_{C(\mathrm{rms})}$, across the capacitor and $V_{RS(\mathrm{rms})}$ for each frequency in Table 38.1. Record these values in Table 38.1.

> Note: Remember to verify the magnitude of the source voltage at each frequency increment.

TABLE 38.1 Capacitive Reactance Measured and Calculated Values

Frequency	*Predicted* X_C	*Measured* $V_{C(PP)}$	*Measured* $V_{RS(PP)}$	*Calculated* $I_{C(PP)}$	*Calculated* X_C	*% Error*
25 Hz						
50 Hz						
75 Hz						
100 Hz						
125 Hz						

6. Using the measured values of $V_{R(\text{rms})}$ and the measured value of $R_{S(\text{rms})}$, calculate the values of the series current $I_{C(\text{rms})}$ for each frequency in Table 38.1. Record these values in Table 38.1.
7. Using the measured values $V_{C(\text{rms})}$ and $I_{C(\text{rms})}$ of the circuit, calculate the capacitive reactance (X_C) of the capacitor at each frequency in Table 38.1. Record these values in Table 38.1.

QUESTIONS

1. As the frequency increased in Table 38.1, what did you observe about the voltage drop across the capacitor? ______________________________
2. Using the measured values of voltage and current, were the predicted values of capacitive reactance equal to the measured values? ____________________
3. Write a statement that describes the relationship between capacitive reactance and an increasing frequency. __________________________________

   ____
4. Take the measured values of X_C from Table 38.1 and plot them in the graph of Figure 38.2. Assume that the capacitive reactance is infinite for a frequency of 0 Hz.
5. The slope of the curve in Figure 38.2 is negative. Explain why.

   ____

   ____

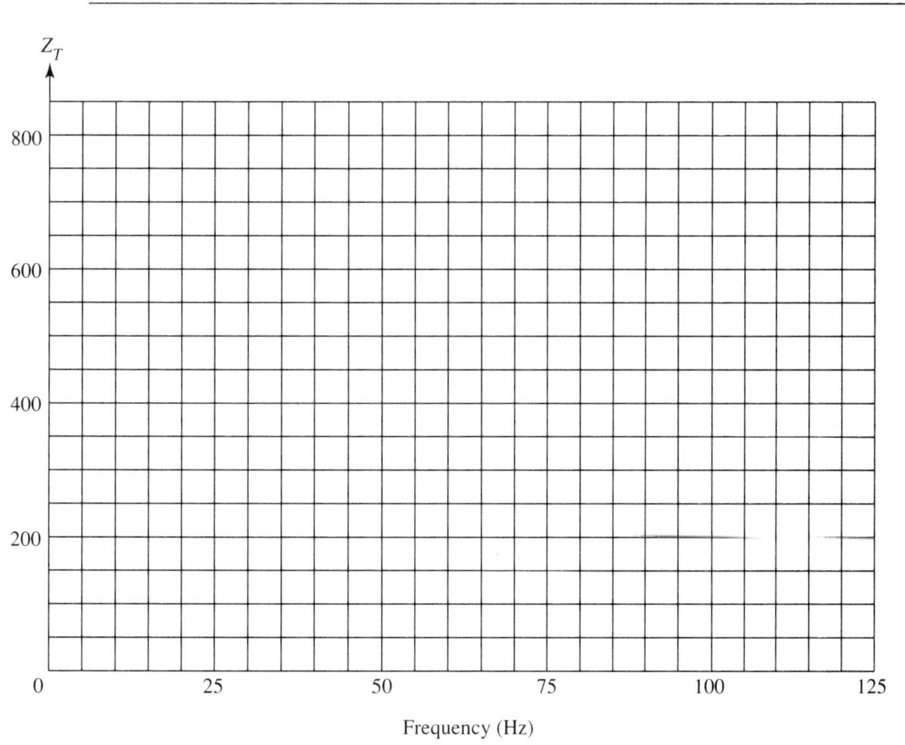

FIGURE 38.2 Reactance versus frequency.

Exercise 39

Series RC Circuit Characteristics

OBJECTIVE

After completing this exercise, you should be able to:

1. Demonstrate that the geometric sum of the voltage drops across V_R and V_C is always equal to the source voltage as determined by Kirchhoff's voltage law.

LAB PREPARATION

Refer to Section 17.1 of *Introductory Electric Circuits.*

DISCUSSION

Series *RC* circuits have many traits that are similar to those of series *RL* circuits. For example, the magnitude of the source voltage equals the geometric sum of the resistor and capacitor voltages. The total impedance in a series *RC* circuit equals the geometric sum of the resistance and capacitive reactance. Both of these relationships are identical to those of series *RL* circuits.

One major difference between *RC* and *RL* circuits is the phase relationship between the source voltage and the circuit current. In a series *RL* circuit, the source voltage leads the circuit current. In series *RC* circuits, the opposite is true: the circuit current leads the source voltage. In either type of circuit, the phase angle between the circuit current and the source voltage depends on the circuit values of resistance and reactance.

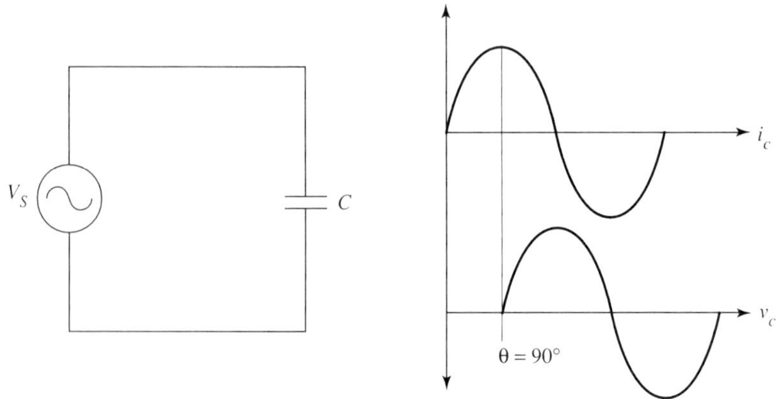

FIGURE 39.1 Phase relationship between capacitor voltage and current.

In this exercise, we will examine the voltage and impedance characteristics of a series *RC* circuit. Two relationships that you may find helpful are:

$$V_S = \sqrt{V_C^2 + V_R^2} \qquad Z = \sqrt{X_C^2 + R^2}$$

MATERIALS

1 oscilloscope
1 function generator
1 DMM
1 150 Ω, 0.5 W resistor
1 10 μf capacitor
1 protoboard
2 sets of test leads

PROCEDURE

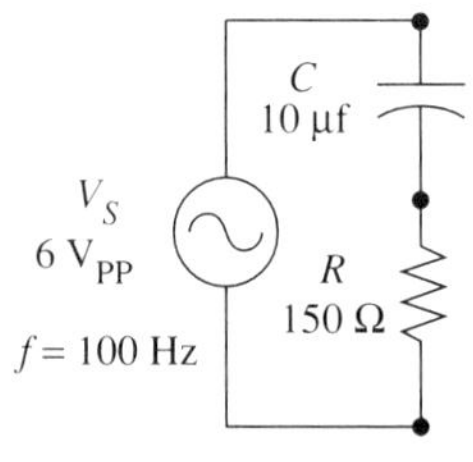

FIGURE 39.2 *RC* test circuit.

1. Measure the ohmic value of R. R = ______
2. Construct the circuit in Figure 39.2. Set the source voltage to 6 V_{PP} at a frequency of 100 Hz.
3. Using the oscilloscope, measure $V_{R(PP)}$ and the phase shift between V_S and V_R.

   V_R = ______ T_C = ______

   t = ______ θ_1 = ______
4. Reverse the position of the capacitor and the resistor. Measure $V_{C(PP)}$ and the phase shift between V_S and V_C.

   $V_{C(PP)}$ = ______ T_C = ______

   t = ______ θ_2 = ______
5. Calculate the total impedance, Z_T, using X_C and R.

   Z_T = ______
6. Calculate I_T using the values of V_S and R. I_T = ______

7. Calculate the total impedance using the source voltage and the circuit current. Compare your answer to the one in step 5. Are the two impedances equal? What is the percent of error between the two values? Show your work.

   Percent of error ________

8. Calculate the geometric sum of V_C and V_R (steps 3 and 4). Does your result equal the source voltage?
9. Does the sum of the two phase angles equal 90°?

   $\theta_1 + \theta_2 =$ ________
10. Use a vector diagram to show in Figure 39.3 that the geometric sum of the two component voltage drops equals the source voltage.

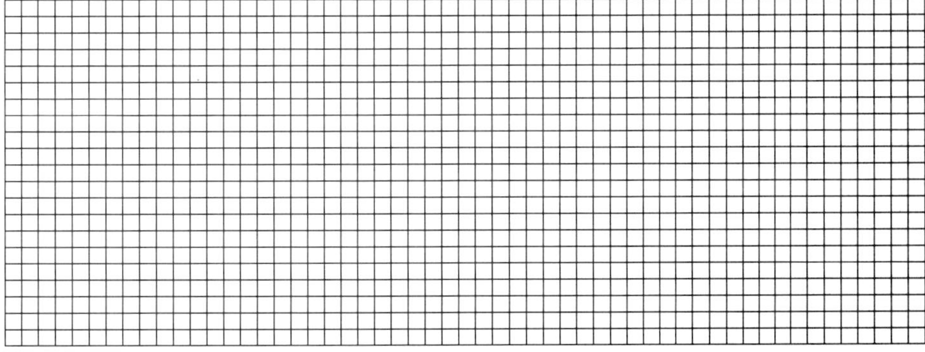

FIGURE 39.3 Phasor diagram of the circuit voltages.

Exercise 40

Series RC Circuit Frequency Response

OBJECTIVES

After completing this exercise, you should be able to:

1. Plot the voltage-versus-frequency curve for a series *RC* circuit.
2. Plot the impedance-versus-frequency curve for a series *RC* circuit.

LAB PREPARATION

Review Section 17.1 of *Introductory Electric Circuits.*

DISCUSSION

For a series ac circuit, the voltage drop across a particular element is directly related to its impedance compared with other series components. Since the impedance of a capacitor changes with frequency, the voltage drops across the *R* and *C* elements will change with a change in frequency. Here are two equations that might be useful:

$$I_T = \frac{V_S}{Z_T} \qquad \theta = \tan^{-1}\frac{V_C}{V_R}$$

TABLE 40.1 ***RC*** **Circuit Value Predictions**

Frequency	*Predicted* X_C	*Predicted* Z_T	*Predicted* I_T	*Predicted* V_C	*Predicted* V_R	*Predicted* $Z_T(\theta)$
1 kHz						
2 kHz						
3 kHz						
4 kHz						
5 kHz						
6 kHz						
7 kHz						
8 kHz						

MATERIALS

1 oscilloscope
1 function generator
1 DMM
1 1 kΩ, 0.5 W resistor
1 0.1 μf capacitor
1 protoboard
2 sets of test leads

PROCEDURE

1. Measure the 1 kΩ resistor. R = ____________________
2. Perform the calculations needed to complete Table 40.1 for the circuit in Figure 40.1.
3. Construct the circuit in Figure 40.1.
4. Set the source voltage, V_S, to 6 V_{PP} at a frequency of 1 kHz.

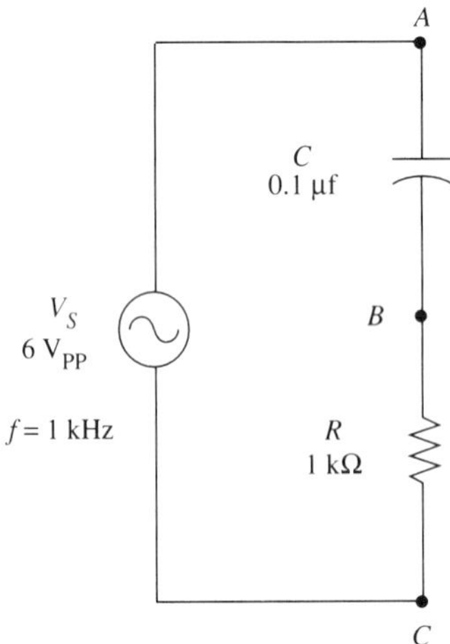

FIGURE 40.1 *RC* test circuit.

TABLE 40.2 Measured and Calculated *RC* Circuit Values

Frequency	*Measured* V_C	*Measured* V_R	*Calculated* I_C	*Calculated* $Z_T\angle\theta$
1 kHz				
2 kHz				
3 kHz				
4 kHz				
5 kHz				
6 kHz				
7 kHz				
8 kHz				

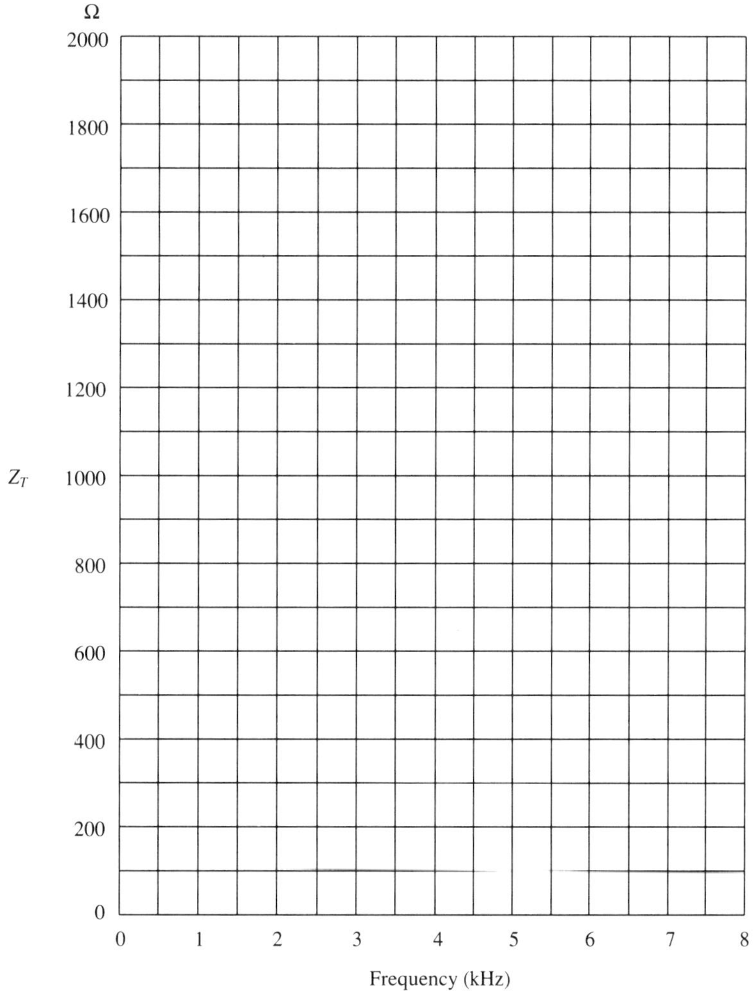

FIGURE 40.2 Impedance versus frequency curve.

5. Measure the values of V_C and V_R for each frequency listed in Table 40.2. Record all these measured values in the table.
6. Calculate the value of the series current (I_C) for each frequency. Calculate the circuit impedance (Z_T) and phase angle (θ) for each frequency. Record all these values in Table 40.2.

QUESTIONS

1. How do the predicted values for the circuit in Figure 40.1 compare to the measured values in Table 40.2?
2. Plot the graph in Figure 40.2, using the values for total impedance, Z_T, from Table 40.2 for each frequency.
3. Plot the graph in Figure 40.3 for V_C and V_R from the values in Table 40.2 for each frequency.
4. From the graph in Figure 40.3, what can you say about Kirchhoff's voltage law? ____

____

____

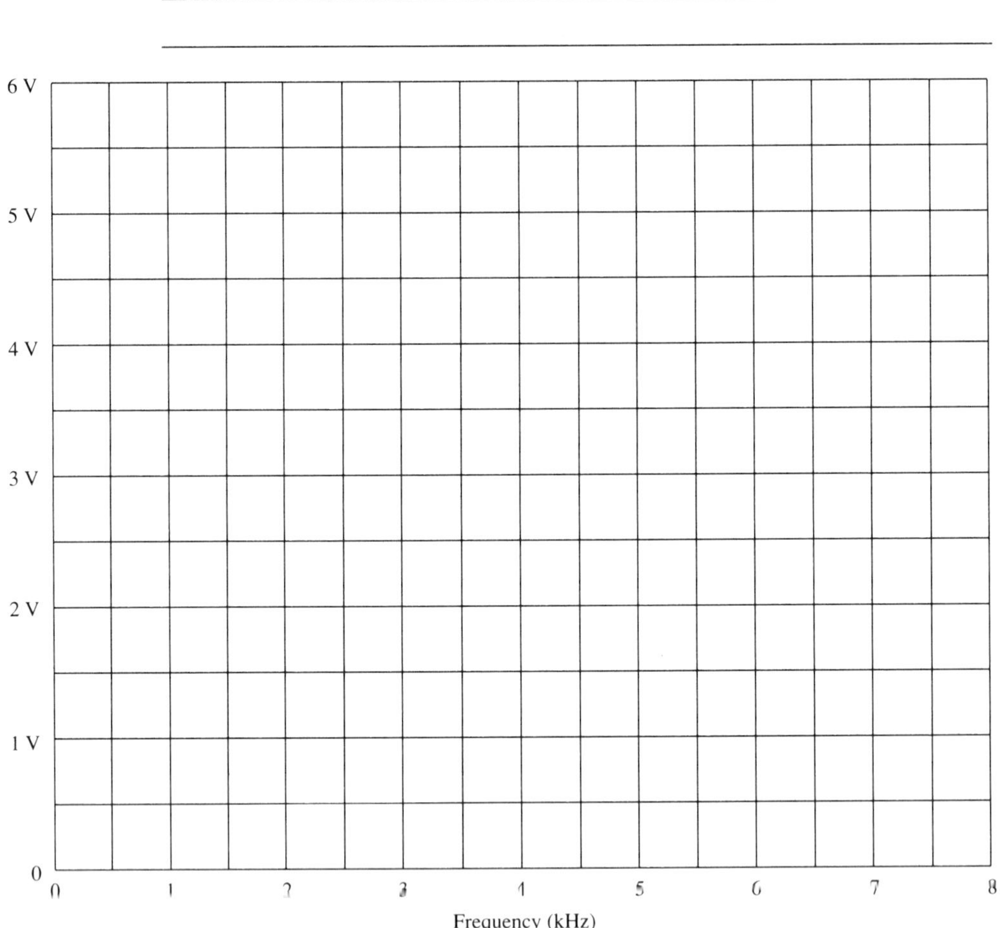

FIGURE 40.3 V_C and V_R versus frequency.

Exercise 41

Parallel RC *Circuit Frequency Response*

OBJECTIVE

After completing this exercise, you should be able to:

1. Apply Kirchhoff's current law to a parallel *RC* circuit.

LAB PREPARATION

Review Section 17.3 of *Introductory Electric Circuits.*

DISCUSSION

In this exercise, you will look at Kirchhoff's current law for parallel *RC* circuits. Kirchhoff's current law states that the geometric sum of the phasor currents entering and leaving a node is equal to zero. All the pertinent equations that you will need are in Section 17.3 of the textbook.

MATERIALS

1 oscilloscope
1 function generator
1 DMM
1 10 Ω, 0.5 W resistor

1 470 Ω, 0.5 W resistor
1 1 μf capacitor
1 protoboard
2 sets of test leads

PROCEDURE

TABLE 41.1 Measured Resistors

Resistance	*Measured Valu e*
$R_S = 10\ \Omega$	
$R = 470\ \Omega$	

1. Measure and record the values listed in Table 41.1.
2. Perform the calculations needed to complete the "Predicted Values" columns in Table 41.2.
3. Construct the circuit in Figure 41.1. Set the frequency generator for 200 Hz and $V_S = 6\ V_{PP}$.
4. Determine the current, $I_{C(PP)}$, in the capacitive branch of the test circuit using the current-sensing resistor (R_s). Record this value of I_C in Table 41.2.
5. Using I_C and V_S, compute the value of X_C. Record this value in Table 41.2.
6. Measure the current, $I_{R(PP)}$ through the 470 Ω resistor and record this value in Table 41.2.
7. Construct the test circuit in Figure 41.2. Measure the total circuit current, $I_{T(PP)}$, using the sensing resistor (R_S). Record this value in Table 41.2.
8. Determine the value of θ by measuring the phase difference between the source voltage (V_S) and the source current (I_T) from the test circuit in Figure 41.2. Record this angle in Table 41.2.

TABLE 41.2 Predicted and Measured Values for the circuit in Figure 41.1.

Variables	*Predicted Values at 200 Hz*	*Measured Values at 200 Hz*	*Predicted Values at 500 Hz*	*Measured Values at 500 Hz*	*Effect on Variable*
X_C					
$I_{C(PP)}$					
$I_{R(PP)}$					
$I_{T(PP)}$					
θ					
Z_T					
P_X					
P_R					
$P_{(APP)}$					

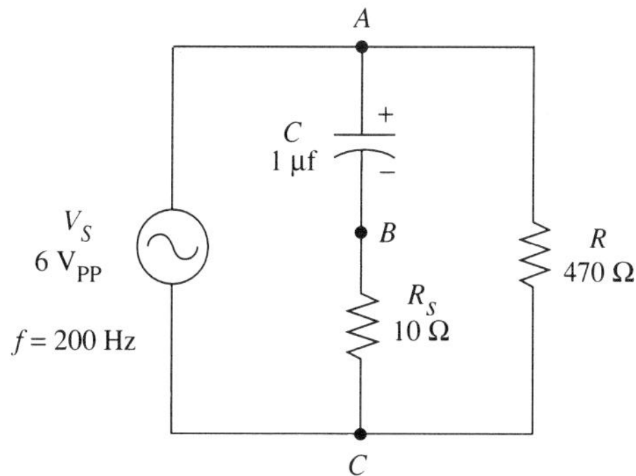

FIGURE 41.1 Parallel *RC* test circuit #1.

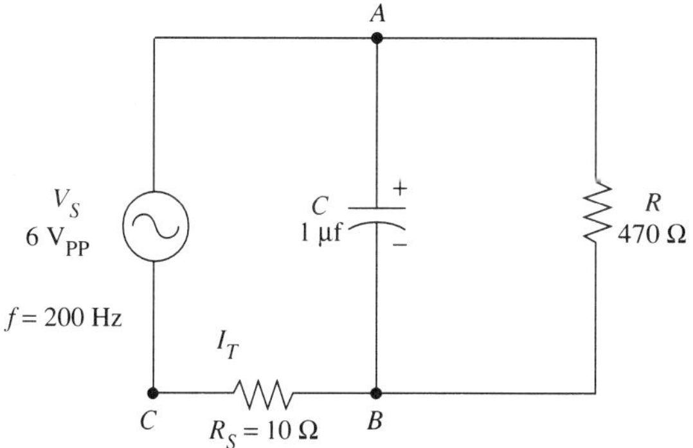

FIGURE 41.2 Parallel *RC* test circuit #2.

9. Using your measured values in Table 41.2, calculate the measured values of Z_T, P_X, P_R, and $P_{(APP)}$. Record these values in Table 41.2.
10. Repeat Steps 3 through 9 with the frequency set to 500 Hz. Record all the measured values in Table 41.2.
11. Determine how each variable is affected when the frequency is changed from 200 Hz to 500 Hz in Table 41.2.

QUESTIONS

1. Does the geometric sum of the currents, I_R and I_C, equal the magnitude of the source current, I_S? ____________________

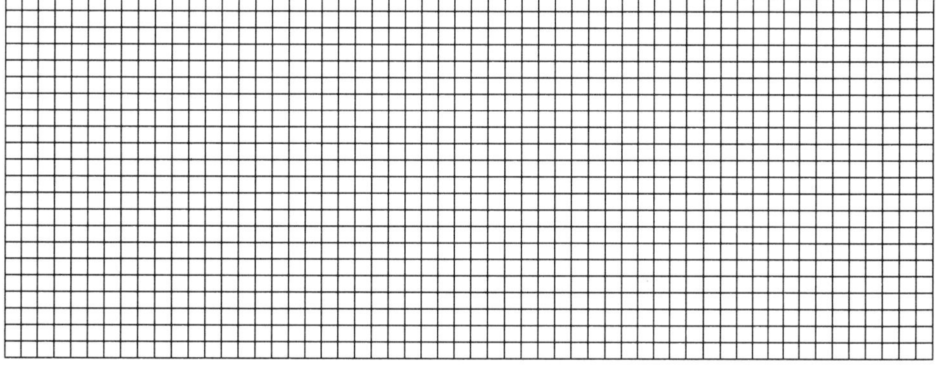

FIGURE 41.3 Phasor diagram of the circuit currents.

2. Using the source voltage as your reference angle, sketch the vector angles of I_R and I_C, and verify that the sum of the two component currents is equal to the source current, I_T. Use the graph in Figure 41.3 for your sketch.
3. State Kirchhoff's current law in terms of your measured results.

____

____

Exercise 42

RC *Transient Response*

OBJECTIVES

After completing this exercise, you should be able to:

1. Determine the steady state response of an *RC* circuit to a dc voltage source.
2. Examine the transient behavior of an *RC* circuit in terms of its voltage and current response to a square wave input voltage source.

LAB PREPARATION

Review Section 17.5 of *Introductory Electric Circuits.*

DISCUSSION

You first encountered transient responses in Exercise 35, where an inductor in an *RL* series circuit opposes any change in circuit current. In an *RC* series circuit, the capacitor opposes any change in circuit voltage. In an *RC* series circuit, when the switch is closed, the voltage drop across the capacitor is initially 0 V. As the time interval increases from 0 s, the voltage across the capacitor increases to its maximum value. When the capacitor voltage is at full charge, the series circuit current is theoretically zero, and the fully charged capacitor acts like an open circuit. As with the *RL* circuit, the current and voltage of the resistor in the *RC* circuit will track the same. At $t = 0$, both I_R and V_R will be at their maximum values. After five time

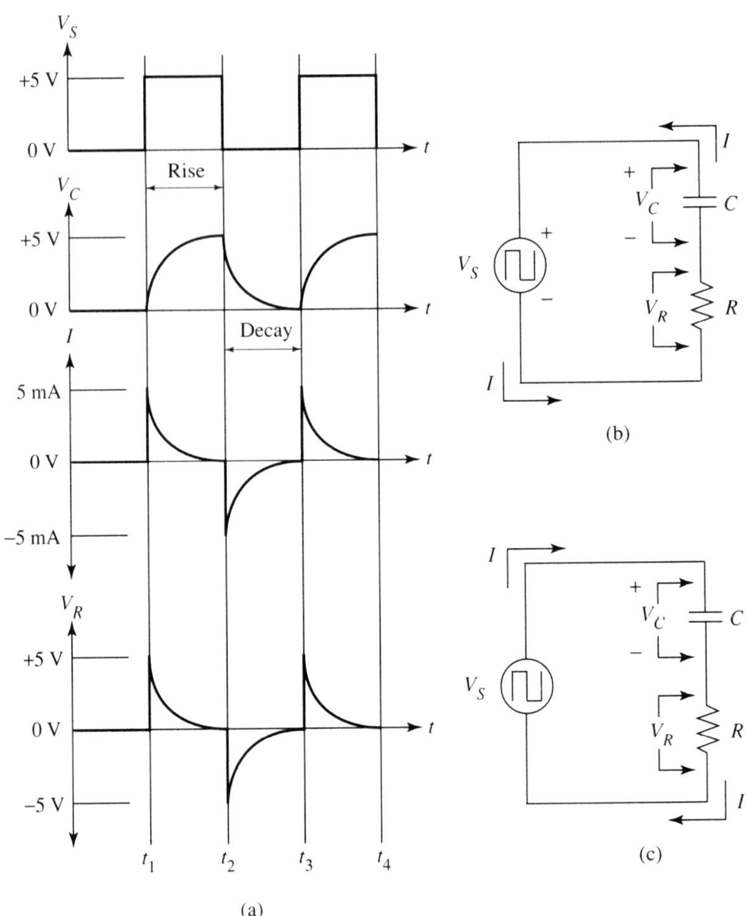

FIGURE 42.1 *RC* circuit waveforms.

constants, I_R and V_R will be at their minimum values. Figure 42.1 gives you a look at the waveforms of an *RC* network. The equations that you will need are in Section 17.5 of the textbook on *RC* time constant. You will also need the universal curve equations.

MATERIALS

1 oscilloscope
1 function generator
1 DMM
1 dc power source
1 10 Ω, 0.5 W resistor
1 122 Ω, 0.5 W resistor
1 1.5 MΩ, 0.5 W resistor
1 1 μf capacitor
1 10 μf capacitor
1 protoboard
2 sets of test leads

TABLE 42.1 Measured Resistance

Resistance	*Measured*
10 Ω	
122 Ω	
1.5 MΩ	

TABLE 42.2 Predicted and Measured Values

Component Values	*Predicted Values*	*Measured Values*
V_C		
V_R		
I		

PROCEDURE

dc Steady State

1. Measure and record the values in Table 42.1.
2. Predict the values indicated in Table 42.2 for the circuit in Figure 42.2. (Assume the capacitor is fully charged.)
3. With the switch open, construct the circuit of Figure 42.2. With the DMM, measure to make sure that the voltage across the capacitor is zero.
4. Close the switch and, with the DMM, measure the voltage drop across V_C and V_R.
5. With the DMM, measure the current I_C through the capacitor. Record these values in Table 42.2.

Transient Response

1. Construct the circuit shown in Figure 42.3.
2. Predict the values called for in Table 42.3.

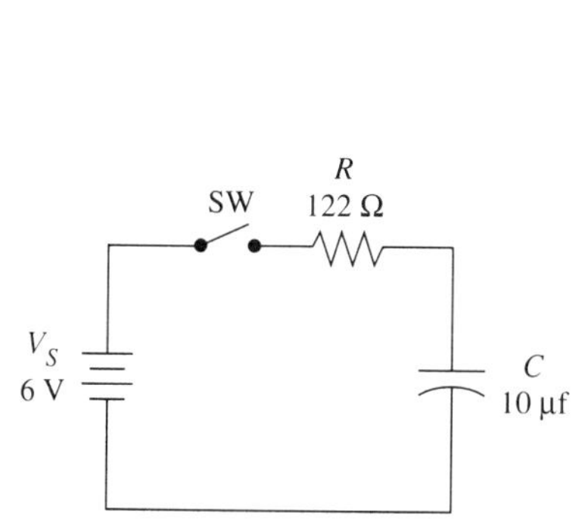

FIGURE 42.2 *RC* steady state test circuit.

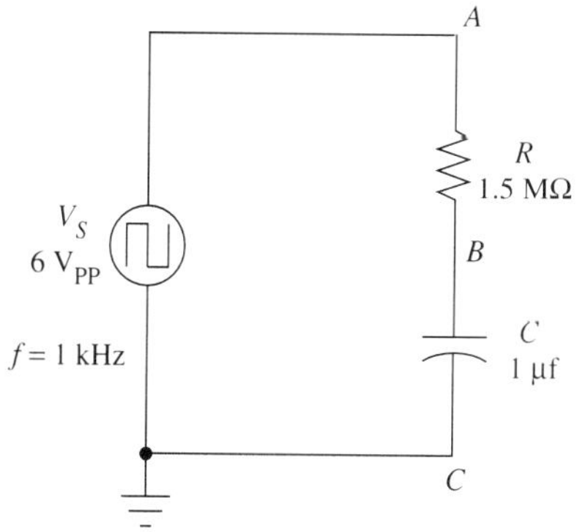

FIGURE 42.3 Transient state *RC* test circuit.

TABLE 42.3 Predicted *RC* Transient Values

Circuit Predictions	0τ	1τ	2τ	3τ	4τ	5τ
V_C						
V_R						
I						

TABLE 42.4 Measured *RC* Transient Circuit Values

Measured Values	0τ	1τ	2τ	3τ	4τ	5τ
V_C						
V_R						
I						

3. Set the function generator frequency for a 1 kHz, 6 V_{PP} square wave output.
4. Set up the oscilloscope to give you one complete cycle of the waveform. Set the volts/div to 1 volt/div. The input again will be set to dc.
5. With the time base control, adjust the display so that the pulse width is 5 major divisions wide. The full 5 time constants of the pulse width are displayed in Figure 42.4.
6. Identify the first five time constants on the display grid. From the waveforms, measure the values of V_C and V_R for each time constant. After you have measured V_C for each time constant, interchange the capacitor and the resistor and measure V_R for each time constant. Record all these values in Table 42.4.
7. From the measured values of V_R, calculate the measured values of I_C for each time constant. Record these values in Table 42.4.

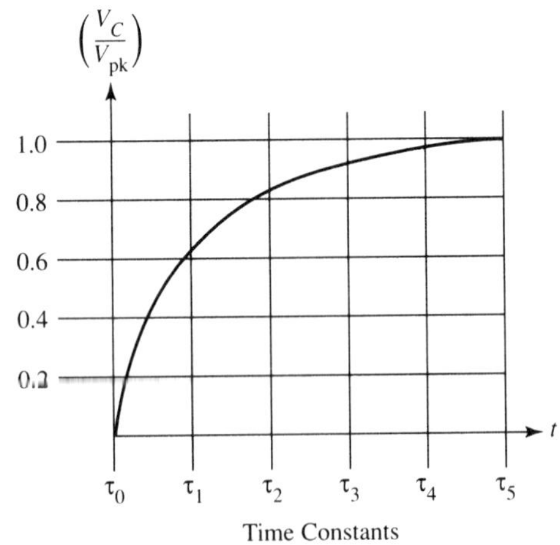

FIGURE 42.4 The rise portion of the universal curve.

QUESTIONS

1. Refer to Table 42.4. Are the measured values close to what you had predicted in Table 42.3? ______
2. What is the value for the time constant (τ) for the circuit of Figure 42.3?

   τ = ______
3. How long did it take for the voltage of the circuit of Figure 42.3 to reach its steady-state value? ______
4. From Table 42.4, graph the response curves of the capacitor voltage, V_C, and the resistor voltage, V_R, for the first five time constants. Use the graph in Figure 42.5.

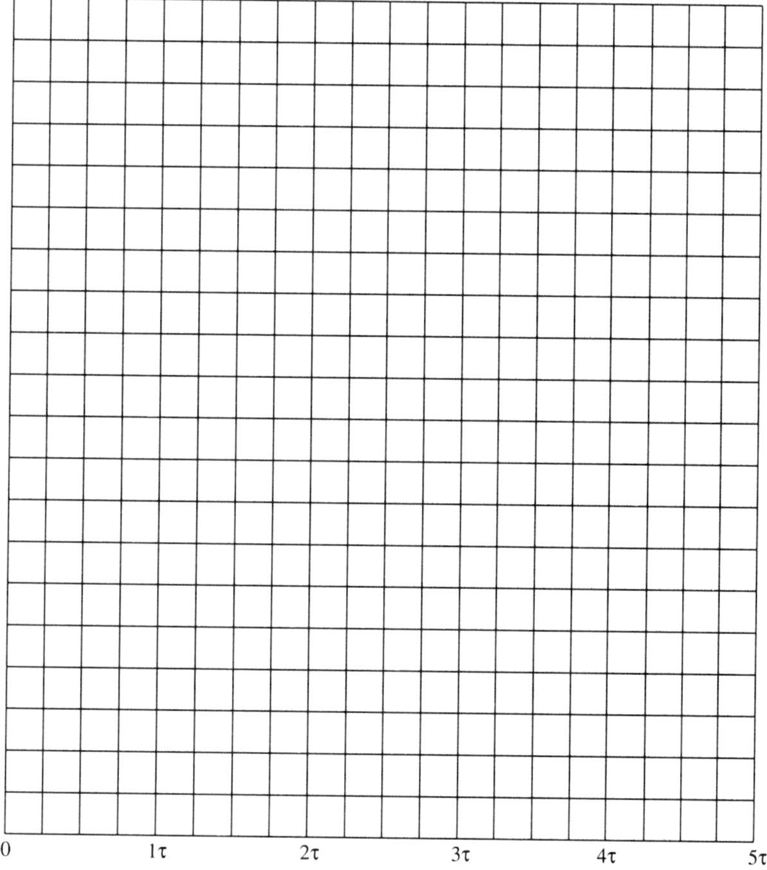

FIGURE 42.5 V_C and V_R transient charge curves.

Exercise 43

Series Resonant Circuit Measurements

OBJECTIVES

After completing this exercise, you should be able to:

1. Calculate and measure the resonant frequency of a series resonant circuit.
2. Analyze the characteristics of a series resonant circuit.

LAB PREPARATION

Review Section 18.3 of *Introductory Electric Circuits*.

DISCUSSION

Resonance Frequency (f_r)

The equation given below identifies the frequency at which inductive and capacitive reactances are equal in magnitude, but their ohmic values are 180° out of phase with each other. This really means that the only resistance in the circuit is any resistance in series with the capacitor and the inductor. Therefore, the circuit current, at resonance, is in phase with the voltage source because the source sees the load as a resistive load.

$$f_r = \frac{1}{2\pi\sqrt{LC}}$$

Below and Above the Resonant Frequency (f_r)

As the frequency increases from 0 Hz toward resonance, the value of X_C is greater than X_L. When this is true, the source sees the circuit as capacitive. In this case, circuit current leads the source voltage by some magnitude of θ. When the frequency is above f_r, the circuit is inductive. In this case, the circuit current lags the source voltage by some magnitude of θ.

MATERIALS

1 oscilloscope
1 function generator
1 DMM
1 33 Ω, 0.5 W resistor
1 10 mH inductor
1 0.1 μf, 50 WVDC capacitor
1 protoboard
2 sets of test leads

PROCEDURE

Resonant Frequency (f_r)

1. Measure the 33 Ω series resistance and the internal resistance of the inductor.

   R_S = ______________

   R_w = ______________
2. Construct the circuit in Figure 43.1. V_S will be set at 2 V_{PP} and the frequency will be set at 1 kHz.
3. Predict the listed values in Table 43.1 and record them in the table.
4. While observing V_{RS}, increase the function generator output frequency until the voltage across the series resistor reaches a maximum value. Stop increasing the frequency and, with the oscilloscope, determine the frequency of the function generator. This value is the resonant frequency (f_r) of the circuit. Record this resonant frequency value in Table 43.1.
5. Using the voltage across the series resistor, determine the magnitude of the circuit current, I_T. Convert this peak-to-peak value to rms. Record this value in Table 43.1.

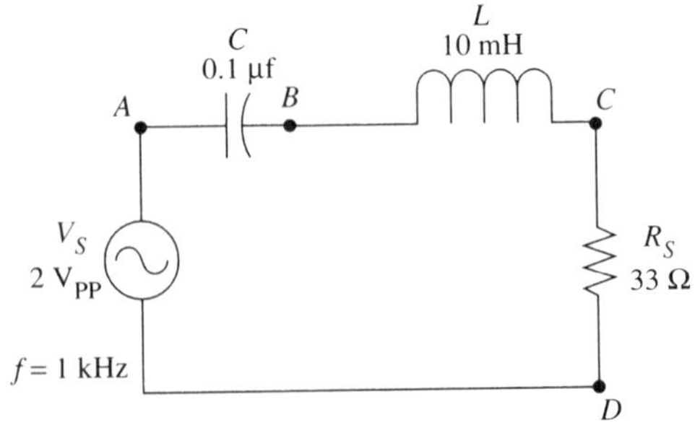

FIGURE 43.1 Series resonance test circuit.

TABLE 43.1 Predicted and Measured Resonant Values

Quantity	*Predicted Values*	*Measured Values*
f_r		
X_L		
X_C		
V_{RS}		
I_S		
V_C		
V_L		

6. Using the DMM, measure the voltage across the capacitor. V_C = ________

   Now measure the voltage across the inductor. V_L = ________
7. Using the voltages across the capacitor and the inductor, calculate the value of the reactance of both components. Record these values in Table 43.1.

I_S Below Resonance

8. Using the same circuit, decrease the frequency from resonance by 1 kHz. The voltage source is still at 2 V_{PP}.
9. Predict the listed quantities in Table 43.2.
10. With the oscilloscope, measure the magnitudes and the phase difference between V_{RS} and the source voltage, V_S. From V_{RS} determine the source current, I_S. Determine the sign of the phase angle. Record these values in Table 43.2.

I_S Above Resonance

11. Increase the frequency to 1 kHz above the resonance frequency. The source voltage is still set to 2 V_{PP}.
12. Predict the values listed in Table 43.3 and record them.
13. With the oscilloscope, measure the magnitudes and the phase difference between V_{RS} and the source voltage, V_S. From V_{RS}, determine the sign of the phase angle. Record these values in Table 43.3.

TABLE 43.2 Measured Values Below the Resonant Frequency

Quantity	*Predicted Values*	*Measured Values*
V_{RS}		
V_S		
I_S		
θ		

TABLE 43.3 Measured Values Above the Resonant Frequency

Quantity	*Predicted Values*	*Measured Values*
V_{RS}		
V_S		
I_S		
θ		

QUESTIONS

Resonance (f_r)

1. Is the measured value of f_r equal to the calculated value of the resonant frequency? If not, what could cause the difference? ________________

   ____

2. Since the inductor contains some resistance, what is the total resistance of the circuit? R_T = ________________

3. Does the current level reflect the addition of R_1 to R_S? Why? ________________

   ____

4. Did the measured values of X_L and X_C come close to the calculated values of these components? ________________

   X_L percent of error ________________

   X_C percent of error ________________

5. Did the current, I_S, below the resonance frequency have a negative phase angle? Why? ________________

   ____

6. Did the current, I_S, above the resonant frequency have a positive phase angle? Why? ________________

   ____

7. The current, I_S, was at its maximum value at the resonant frequency, and I_S decreased in magnitude above and below resonance frequency. What does this tell you about the impedance of the series resonant circuit as it approaches resonance and after it passes its resonant frequency? ________________

   ____

   ____

Exercise 44

Parallel Resonant Circuit Measurements

OBJECTIVES

After completing this exercise, you should be able to:

1. Calculate and measure the resonant frequency of a parallel resonant circuit.
2. Analyze the current characteristics of a parallel resonant circuit.

LAB PREPARATION

Review Section 18.3 of *Introductory Electric Circuits.*

DISCUSSION

Resonant Frequency (f_r)

In a parallel resonant circuit, the reactances X_L and X_C are equal in magnitude but 180° out of phase. The source voltage, V_S, is applied to both branch reactances. The difference is that at resonance, the currents through both parallel branches are equal in magnitude but 180° out of phase with each other. Therefore, in the reactive branches, the net current is equal to 0 A. Since the net current through the parallel resonant circuit is 0 A, the parallel resonant circuit acts like an open circuit. For all practical purposes, the total parallel resonant reactance is infinite and the circuit current, I_S, is effectively equal to 0 A. At 0 Hz, the inductive reactance would be almost 0 Ω, while the capacitive reactance would be at a maximum. The current, I_S, would be at its maximum value. At an extremely high frequency, the capacitive reactance would be very low, but the inductive reactance would be very high. In effect the capacitor's capacitive reactance would act like a near short. Therefore, at the resonant frequency, the source sees an open circuit. The same formula is used for parallel resonant frequency as was used for series resonant frequency.

$$f_r = \frac{1}{2\pi\sqrt{LC}}$$

MATERIALS

1 oscilloscope
1 function generator
1 DMM
2 10 Ω, 0.5 W resistors
1 33 Ω, 0.5 W resistor
1 10 mH inductor
1 0.1 μf, 50 WVDC capacitor
1 protoboard
2 sets of test leads

PROCEDURE

Resonant Frequency (f_r)

1. Measure the series resistor R_S (33 Ω) and the winding resistance of the inductor.

   $R_S =$ ______________

   $R_w =$ ______________
2. Construct the circuit of Figure 44.1 ($V_S = 2\ V_{PP}, f = 1$ kHz).
3. Predict the listed values at the resonant frequency in Table 44.1 and record them there.
4. While observing the voltage across R_S, increase the frequency of the function generator until the voltage across the resistor is at its minimum value. Stop increasing the frequency. Using the oscilloscope, determine the frequency of the function generator. This is the resonant frequency (f_r) of the parallel circuit. Record this value in Table 44.1.
5. Record the voltage, V_{RS}, at this frequency in Table 44.1. Using this value, calculate the circuit current and record this value in Table 44.1.

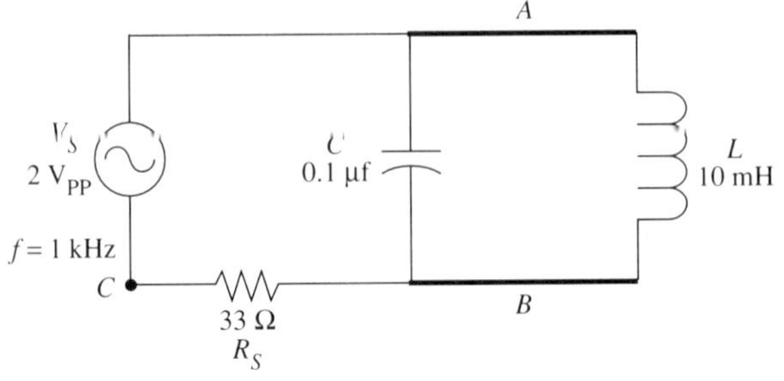

FIGURE 44.1 Parallel resonant test circuit #1.

TABLE 44.1 Predicted and Measured Parallel Circuit Values

Quantity	*Predicted Values*	*Measured Values*
f_r		
X_C		
X_L		
V_{RS}		
I_S		
V_{RS1}		
V_{RS2}		
$I_C \angle\theta$		
$I_L \angle\theta$		
V_C		
V_L		

6. Construct the circuit of Figure 44.2. Use the same voltage and the resonant frequency from Step 4. Measure the two 10 Ω resistors.

   R_{S1} = ____________

   R_{S2} = ____________
7. With V_S set to 2 V_{PP} and the function generator set to the resonant frequency, measure the voltage drop across V_{RS1}. Record this value in Table 44.1.
8. Calculate the current, I_C, through R_{S1} and its phase shift with respect to the voltage drop, V_C, across the whole capacitor branch. Record the magnitude and the phase angle of the current with its sign in Table 44.1.
9. Using I_C and V_C, calculate the measured value of X_C. Record this value in Table 44.1.
10. Measure the voltage drop across V_{RS2} and record this value in Table 44.1.
11. Calculate the current, I_L, through R_{S2} and its phase shift with respect to the voltage drop, V_L, across the whole inductor branch. Record the magnitude of the current and phase shift with its sign in Table 44.1.
12. Using the values of I_L and V_L, calculate the measured value of X_L. Record this value in Table 44.1.

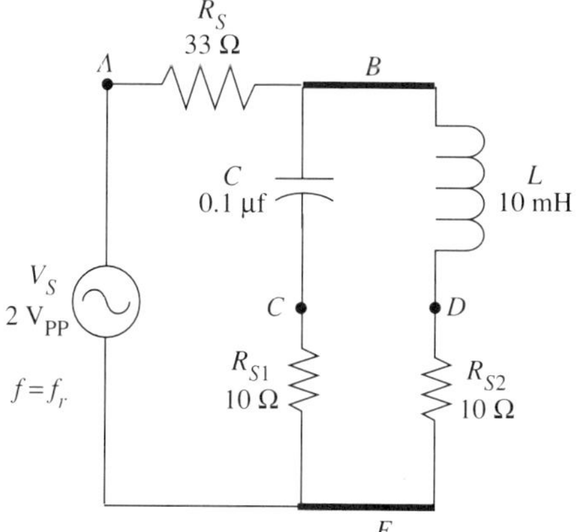

FIGURE 44.2 Parallel resonant test circuit #2.

QUESTIONS

1. Is the measured value of resonant frequency equal to the calculated value of the resonant frequency? If not, what could cause the difference? ____________________

   ____
2. Was the value of the circuit current what you had predicted? If not, what could cause the difference? ____________________
3. From the measured values of I_C and I_L, were the currents of approximately equal magnitude? ____________________
4. Were the measured values of the phase angle of I_C and I_L approximately 180° out of phase with each other? ____________________
5. Given that the current, I_S, is at its maximum value for a series resonant circuit and the current for a parallel resonant circuit is at its minimum value, describe in your own words the difference with respect to the impedance of each type of circuit.

   Series resonance ____________________

   ____

   Parallel resonance ____________________

   ____

Exercise 45

Series RLC *Circuit Measurements*

OBJECTIVE

After completing this exercise, you should be able to:

1. Describe and analyze the operation of any series *RLC* circuit.

LAB PREPARATION

Review Section 18.4 of *Introductory Electric Circuits.*

DISCUSSION

You have analyzed series and parallel *RC* and *RL* circuits, and series and parallel resonant circuits. You will now look at a simple series *RLC* circuit and investigate its characteristics.

MATERIALS

1 oscilloscope
1 function generator
1 DMM
1 100 Ω, 0.5 W resistor
1 0.1 μf capacitor
1 1 mH inductor

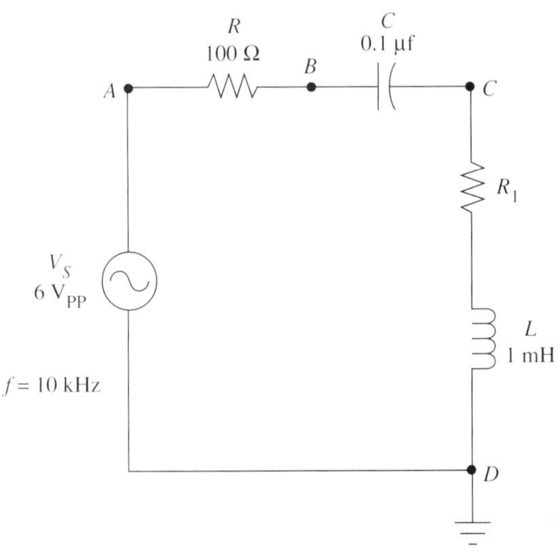

FIGURE 45.1 Series *RLC* test circuit.

1 protoboard
2 sets of test leads

PROCEDURE

1. Measure the resistances R and R_w of the inductor.

   $R =$ ______________

   $R_w =$ ______________
2. Construct the circuit in Figure 45.1.
3. Set the function generator output at 6 V_{PP} and the frequency to 10 kHz.
4. For the quantities listed in Table 45.1, predict the values of the unknowns and record their values.
5. Measure $V_{L(PP)}$, $V_{C(PP)}$, and $V_{R(PP)}$. As you measure each of these voltages, make sure that you interchange first the capacitor with the inductor, then the resistor with the capacitor.

TABLE 45.1 Predicted and Measured Circuit Values

Components	*Predicted Values*	*Measured Values*	*Percent of Error*
$V_{S(PP)}$			
$V_{L(PP)}$			
$V_{C(PP)}$			
$V_{R(PP)}$			
I_S			
Z_T			
X_L			
X_C			

6. Calculate the value of $I_{S(PP)}$ from measured values in Table 45.1.
7. Calculate the value of the circuit impedance, Z_T, from the measured values in Table 45.1.
8. Using the current and the measured voltages across the inductor and the capacitor, calculate the values of X_L and X_C. Record these values in Table 45.1.
9. Using the nominal values of the inductor and the capacitor, calculate the polar impedance, Z_T, for the circuit. Show your work.

   Z_T = ______________

QUESTIONS

1. Is the ohmic value of R_1 large enough to be needed in the circuit calculations? Explain your answer. ______________________________
2. Which of the measured values produced the largest error? Why?

   ____
3. Draw the phasor diagram that will verify Kirchhoff's voltage law by solving for $V_S = \sqrt{V_R^2 + (V_L - V_C)^2}$. Use the values from Table 45.1. Use $V_S = V_S \angle 0°$ as the reference angle. Show your work on the graph in Figure 45.2.
4. Did the predicted value of the impedance equal the measured value of the impedance? What was the percent of error? Why? ______________________________

   ____

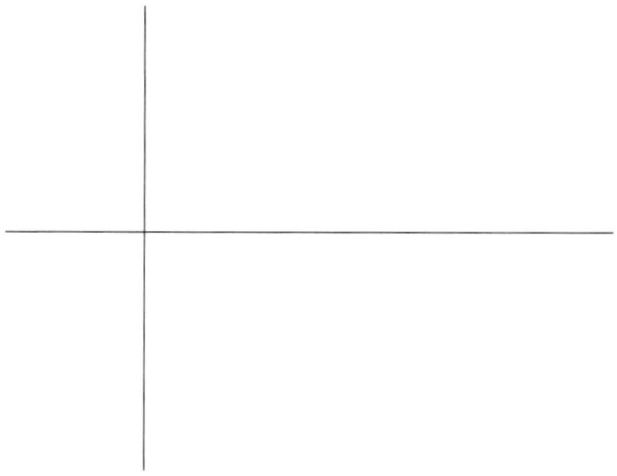

FIGURE 45.2 Kirchhoff's voltage law phasor diagram.

Exercise 46

Parallel RLC *Circuit Measurements*

OBJECTIVE

After completing this exercise, you should be able to:

1. Describe and analyze the operation of any parallel *RLC* circuit.

LAB PREPARATION

Review Section 18.4 of *Introductory Electric Circuits*.

DISCUSSION

Important in the analysis of any parallel *RLC* circuit is the understanding that the voltage is the same across each parallel branch. Also, it is important that the current through the resistor is in phase with the source voltage, and the currents through the inductor and the capacitor are 180° out of phase with each other. All of the methods you learned in parallel *RC* and *RL* circuits hold true here.

$$I_T = \sqrt{I_{LC}^2 + I_R^2} \qquad \theta = \tan^{-1} \frac{I_{LC}}{I_R}$$

TABLE 46.1 Measured Resistances

Resistors	*Measured*
R	
R_1	
R_{S1}	
R_{S2}	
R_{S3}	

MATERIALS

1 oscilloscope
1 function generator
1 DMM
1 1.2 kΩ, 0.5 W resistor
3 10 Ω, 0.5 W resistor
1 0.001 μf capacitor
1 protoboard
2 sets of test leads

PROCEDURE

1. Measure all the resistors listed and record their values in Table 46.1. Make sure that you label each sensing resistor for further identification.
2. Construct the circuit of Figure 46.1 ($V_S = 6\ V_{PP}, f = 10$ kHz).
3. Predict the values listed in Table 46.2 and record them.
4. With the oscilloscope, measure the voltage drops across R, R_{S1}, and R_{S2} and across nodes A and B. Record these voltage readings in Table 46.2.
5. With the oscilloscope, measure the voltage drop across R_{S3} (from node C to B). Record this value in Table 46.2.
6. Using your measured values, calculate I_R, I_C, I_L, and $I_{S/\theta}$. Record these four currents in Table 46.2.
7. Using the calculated currents and the source voltage, determine the values of circuit impedance, $Z \angle\theta°$, and the inductor and capacitor reactances, X_L and X_C. Record these values in Table 46.2.

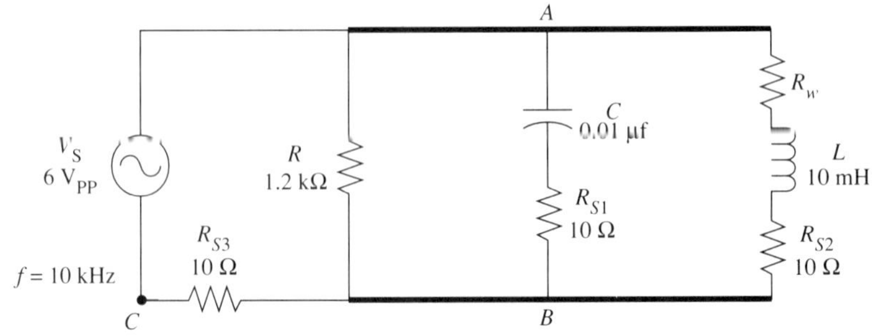

FIGURE 46.1 Parallel *RLC* test circuit.

TABLE 46.2 Predicted and Measured Parallel *RLC* Circuit Values

Components	*Predicted Values*	*Measured Values*	*Percent of Error*
$V_{R(PP)}$			
$V_{R1(PP)}$			
$V_{R2(PP)}$			
$V_{R3(PP)}$			
I_R			
I_C			
I_L			
$I_S \angle\theta$			
Z_T			
X_L			
X_C			

QUESTIONS

1. Which of the measured values gave you the largest error? Why? ____________

   ____

2. Draw the phasor diagram in Figure 46.2 that will verify Kirchhoff's current law by using the equations in the discussion. Remember that $V_S = V_S \angle 0°$, therefore, $I_S = I_S \angle 0°$.
3. Was the predicted value of the impedances equal to the measured value? If not, why not? ____

   ____

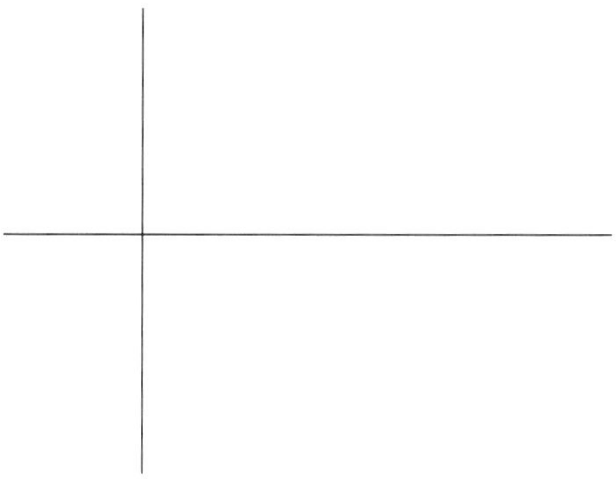

FIGURE 46.2 Kirchhoff's current law phasor diagram.

Exercise 47

Simple Low-Pass and High-Pass Filters

OBJECTIVES

After completing this exercise, you should be able to:

1. Describe and analyze low-pass filter circuit characteristics.
2. Describe and analyze high-pass filter circuit characteristics.

LAB PREPARATION

Review Sections 19.4 and 19.5 of *Introductory Electric Circuits.*

DISCUSSION

Filters are specific circuit networks that manipulate a circuit response to achieve a required output signal. Low-pass and high-pass filters are designed to reject or pass specific ranges of frequencies. These filters use either capacitors or inductors in several different configurations.

Low-Pass Filter

The low-pass filter allows a band of frequencies to pass from 0 Hz to a higher cutoff frequency, f_c, where f_c represents the highest frequency that the circuit will pass from its input to its output. This upper frequency is designated as the point at which the reactance of the inductor or the capacitor is equal to the series resistance.

High-Pass Filter

The high-pass filter blocks a band of frequencies from 0 Hz to a higher frequency, f_C, where f_C represents the lowest frequency that the circuit will pass from its input to its output. This lowest frequency is designated as the point at which the reactance is equal to the series resistance.

General Information

The cutoff frequency for a filter is the frequency at which

$$\frac{V_{out}}{V_{in}} = 0.707$$

At this point, the capacitive reactance is equal to the series resistance ($X_C = R$). This relationship of V_{out}/V_{in} also defines the voltage gain (A_V) of the circuit.

$$f_c = \frac{1}{2\pi RC}$$

where f_c = cutoff frequency

MATERIALS

1 oscilloscope
1 function generator
1 DMM
1 22 kΩ, 0.5 W resistor
1 3.3 nf capacitor
1 protoboard
2 sets of test leads

PROCEDURE

Low-Pass Filter

1. Measure the value of the resistor.

   R = _______________
2. Construct the circuit in Figure 47.1. (V_S = 10 V_{PP}, f = 200 Hz).
3. Predict the cutoff frequency for this circuit.

   f_C = _______________
4. Measure the output voltage for each frequency given in Table 47.1. Record these values in Table 47.1.

> Remember: Every time you change the frequency, you change the load impedance that the source sees. Check that the input voltage is a constant value.

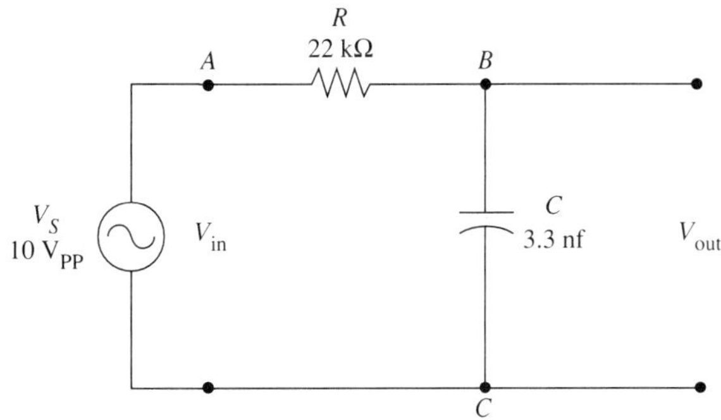

FIGURE 47.1 Low-pass filter test circuit.

TABLE 47.1 Low-Pass Filter Output Voltage Readings

Frequency (kHz)	V_o	$A_V = \frac{V_o}{V_i}$
0.1		
0.2		
0.4		
0.6		
0.8		
1.0		
1.2		
1.4		
1.6		
1.8		
2.0		
3.0		
4.0		
6.0		
8.0		
10.0		

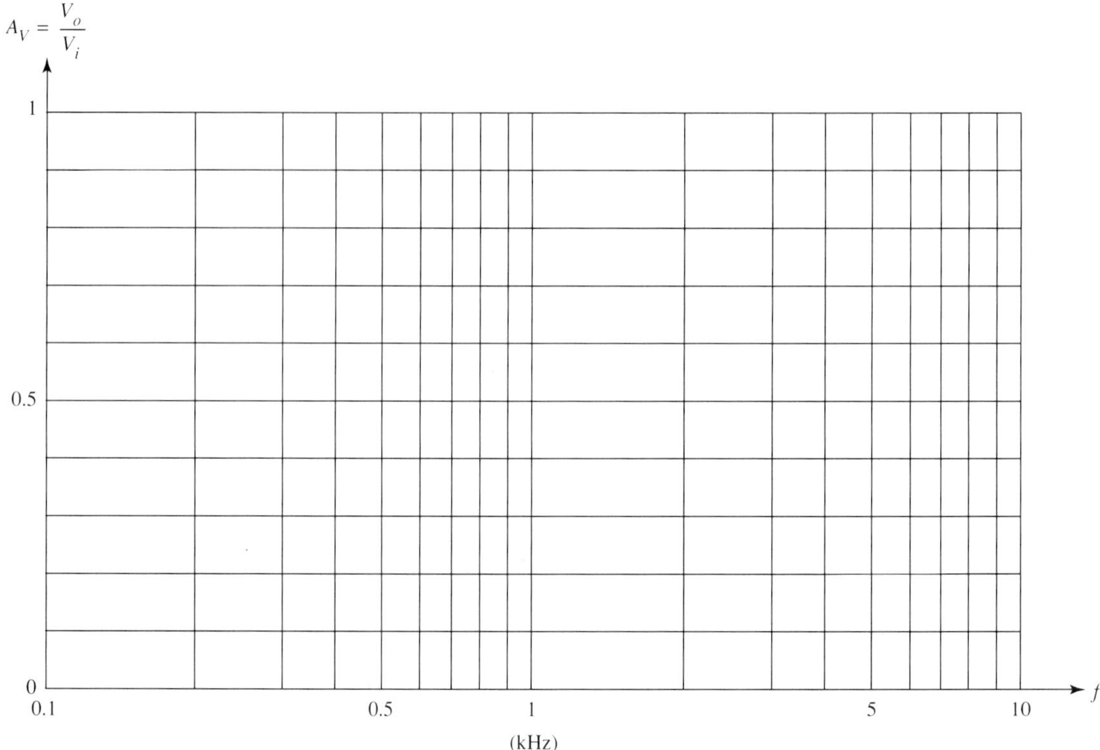

FIGURE 47.2 Frequency response, low-pass filter.

5. Plot the voltage gain, A_V, versus frequency on the graph in Figure 47.2. Mark each point and identify the type of curve.

High-Pass Filter

1. Construct the circuit of Figure 47.3 ($V_S = 10$ V, $f = 200$ Hz).
2. Predict the cutoff frequency for this circuit.

   $f_c =$ _______________
3. Measure the output voltage for each frequency given in Table 47.2.
4. Plot the voltage gain, A_V, versus frequency on the graph in Figure 47.4. Mark each point and identify the type of curve.

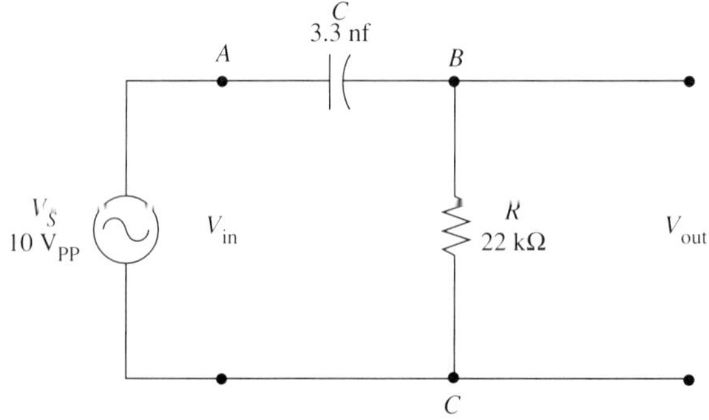

FIGURE 47.3 High-pass filter test circuit.

TABLE 47.2 High-Pass Filter Output Voltage Readings

Frequency (kHz)	V_o	$A_V = \frac{V_o}{V_i}$
0.1		
0.4		
0.8		
1.0		
1.2		
1.6		
1.8		
2.0		
2.2		
2.6		
3.0		
4.0		
6.0		
8.0		
10.0		

QUESTIONS

Low-Pass Filter

1. On the graph in Figure 47.2, lightly sketch the 0.707 A_V horizontal reference line. At the point where the plotted curve crosses that 0.707 line, we have the cutoff frequency for the low-pass filter circuit. Does the measured frequency from the plot compare with the predicted value you found in Step 3? What is the percent of error? ________________
2. From the information seen on the graph, what is the useful range of output frequencies for the filter? ________________

____

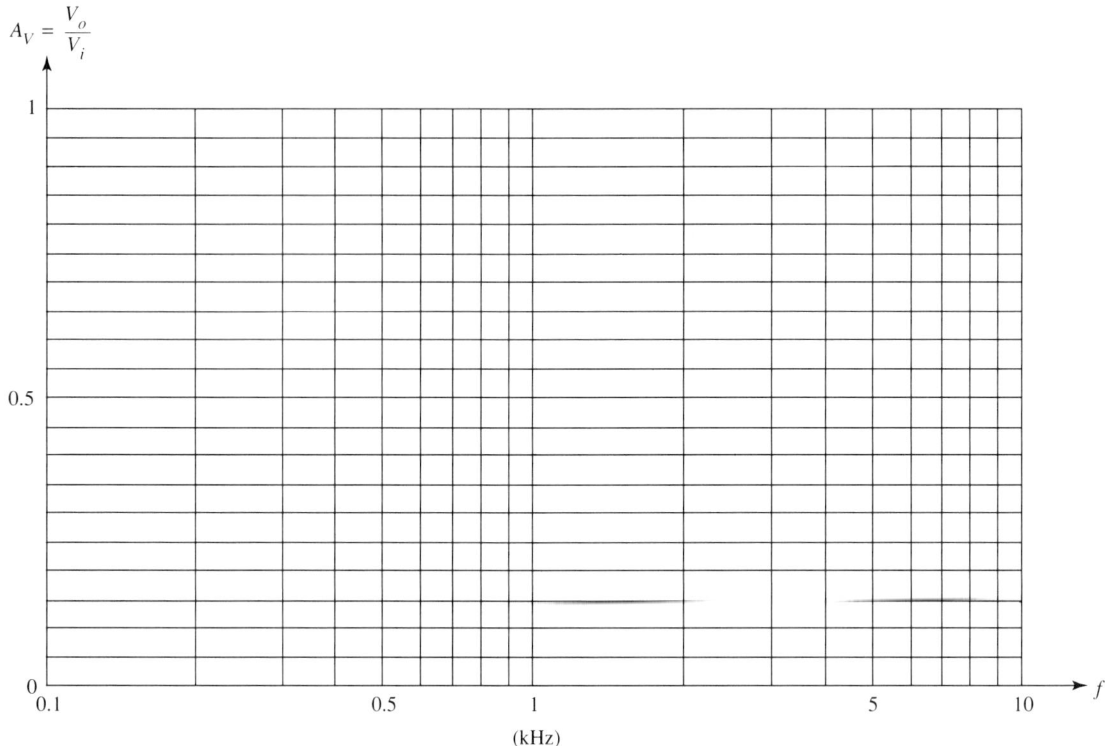

FIGURE 47.4 Frequency response, high-pass filter.

High-Pass Filter

3. On the graph in Figure 47.4, lightly sketch the 0.707 A_V horizontal reference line. At the point where the plotted curve crosses the 0.707 line, we have the cutoff frequency for the high-pass filter circuit. Does the measured frequency from the plot compare with the predicted value found in Step 2? What is the percent of error? ______
4. From the information seen on the graph, what is the useful range of output frequencies for the filter? ______

______

Exercise 48

Simple Tuned Bandpass Filter

OBJECTIVE

After completing this exercise, you should be able to:

1. Analyze the operation of a tuned bandpass filter.

LAB PREPARATION

Review Section 19.6 of *Introductory Electric Circuits*.

DISCUSSION

The operation of a series bandpass filter is easiest to understand when the filter is represented as an equivalent circuit like the one shown in Figure 48.1. In this circuit, the series reactance, X_S, of the filter is represented as a series component between the source and the load. The response curve of a bandpass filter is illustrated in Figure 48.2.

In this exercise, you will determine the response of the filter when X_L is equal and opposite to X_C. This occurs when the filter has reached the state of resonance. You will also investigate the response of the bandpass filter over a significant range of frequencies. Some equations that might be useful to you are:

$$f_s = \frac{1}{2\pi\sqrt{LC}} \qquad Q_L = \frac{X_L}{R_W + R_L} \qquad V_L = \frac{V_i R_L}{R_W + R_L}$$

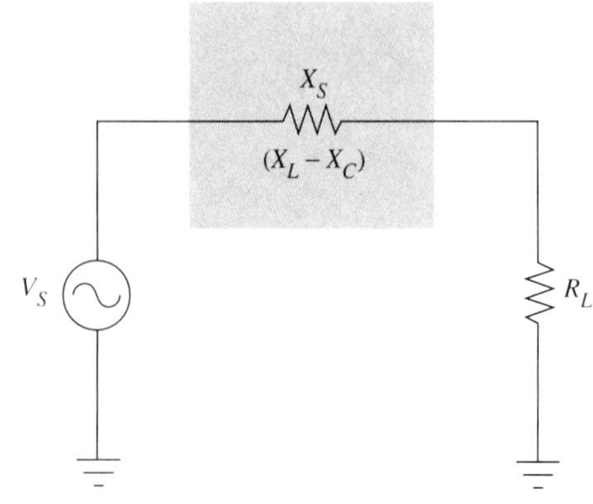

FIGURE 48.1 The reactive equivalent of the filter.

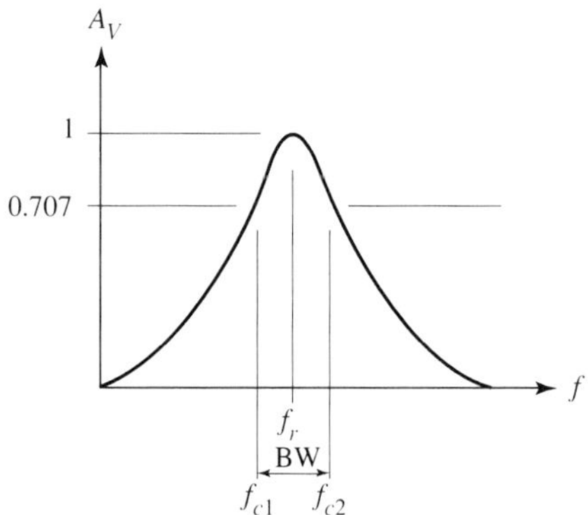

FIGURE 48.2 Response curve of a bandpass filter.

MATERIALS

1 oscilloscope
1 function generator
1 DMM
1 1 mH inductor
1 0.03 μf capacitor
1 56 Ω, 0.5 W resistor
1 protoboard
3 test leads

PROCEDURE

1. Measure the resistor values and record them in Table 48.1.
2. Construct the circuit in Figure 48.3 ($V_S = 6\ V_{PP}$ and the function generator is set to 1 kHz).

TABLE 48.1 Measured Resistance

Resistance	*Measured Resistance*
$R_L = 56\ \Omega$	
R_w	

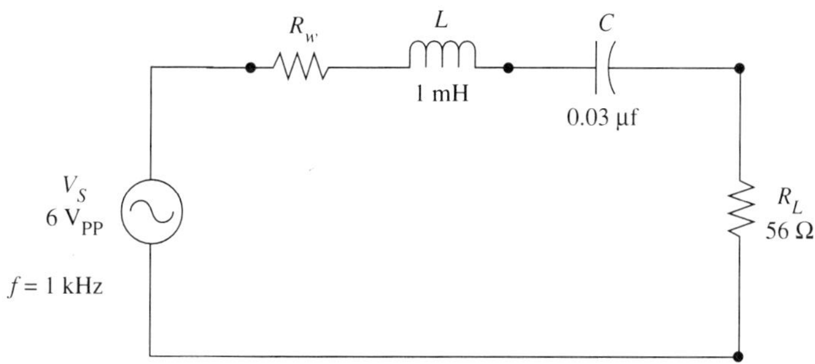

FIGURE 48.3 Bandpass test circuit.

3. Predict and record the circuit values for the bandpass filter at resonance, as indicated in Table 48.2.
4. Measure the values of V_o with the oscilloscope for each frequency indicated in Table 48.3. Record the values of V_o in Table 48.3. Make sure that V_i is 6 V_{PP} for each change of frequency.
5. From the values of V_o, calculate the measured value of the voltage gain, A_V, and record these values in Table 48.3.
6. Plot the voltage gain with respect to the frequency indicated on the graph in Figure 48.4. With a dashed line, identify the 0.707 A_V points on the graph.

TABLE 48.2 Predicted Resonant Frequency Circuit Values

f_S	Q_L	BW	f_{c1}	f_{c2}	$V_L = V_o$	A_V

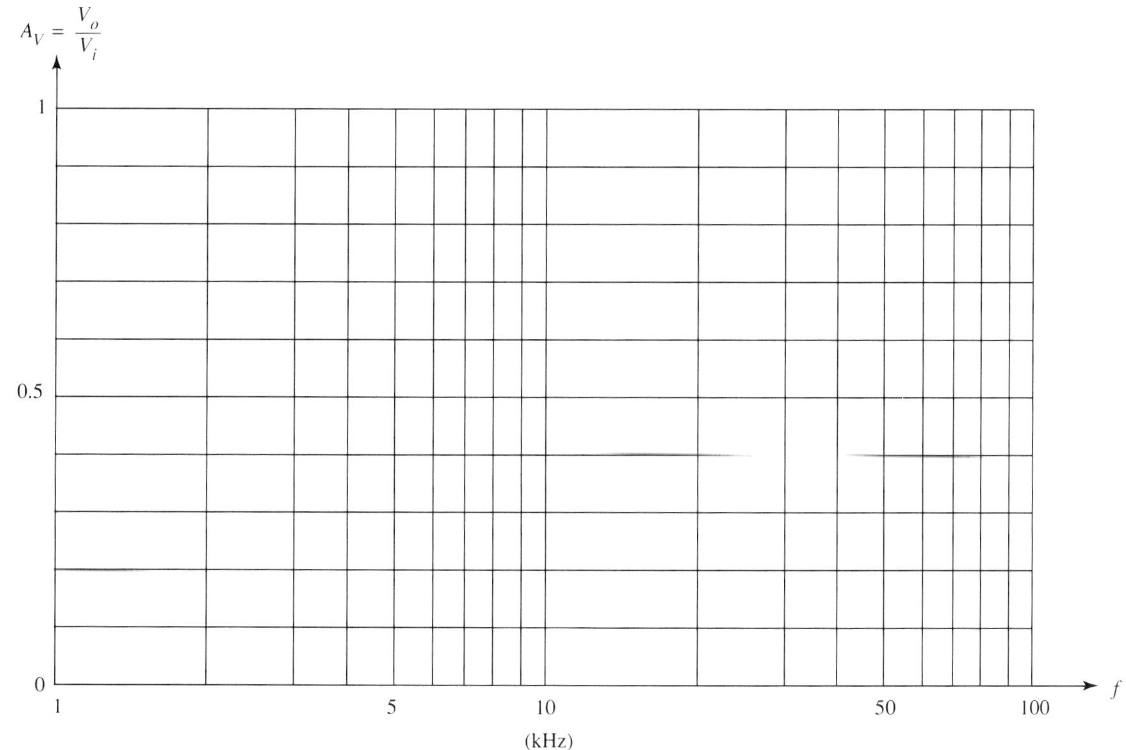

FIGURE 48.4 (kHz) Voltage gain versus frequency response.

TABLE 48.3 Measured Values of V_o and A_V

Test Frequency	V_{oPP}	$A_V = V_o/V_i$
10 kHz		
15 kHz		
20 kHz		
24 kHz		
28 kHz		
29 kHz		
30 kHz		
31 kHz		
32 kHz		
33 kHz		
34 kHz		
35 kHz		
36 kHz		
38 kHz		
40 kHz		
50 kHz		
60 kHz		

QUESTIONS

1. Did the total bandwidth for your circuit come close to what you had calculated? ________________
2. Did the voltage gain come close to what you had calculated in Table 48.3?
   Percent of difference ________________
3. Given the voltage gain for your circuit, were the 0.707 points close to what you expected for f_{c1} and f_{c2}? Explain your answer. ________________

NOTES

NOTES

NOTES

NOTES

NOTES

NOTES

NOTES

NOTES